Advanced Earthquake Engineering

Advanced Earthquake Engineering

Editor: Agnes Nolan

MURPHY & MOORE
www.murphy-moorepublishing.com

Published by Murphy & Moore Publishing
1 Rockefeller Plaza,
New York City, NY 10020, USA
www.murphy-moorepublishing.com

Advanced Earthquake Engineering
Edited by Agnes Nolan

International Standard Book Number: 978-1-63987-010-3 (Hardback)

Cataloging-in-Publication Data

Advanced earthquake engineering / edited by Agnes Nolan.
p. cm.
Includes bibliographical references and index.
ISBN 978-1-63987-010-3
1. Earthquake engineering. 2. Earthquakes. 3. Civil engineering. 4. Engineering geology.
5. Earthquake resistant design. 6. Reinforced concrete construction. I. Nolan, Agnes.
TA654.6 .A38 2022
624.176 2--dc23

Contents

Preface

An interdisciplinary branch of engineering that designs and analyzes structures such as buildings and bridges keeping earthquakes in mind is referred to as earthquake engineering. The main aim of earthquake engineering is to foresee the potential consequences of strong earthquakes on metropolitan areas and civil infrastructure. It also attempts to design, construct and maintain structures in accordance with building codes and compliance. It is considered as a subset of structural engineering, geotechnical engineering, mechanical engineering, chemical engineering and applied physics. Earthquake engineering has expanded its scope to encompass disciplines from the wider field of civil engineering, nuclear engineering, sociology, political science, economics, and finance. This book is a compilation of chapters that discuss the most vital concepts and emerging trends in earthquake engineering. It elucidates new techniques and their applications in a multidisciplinary manner. This book aims to equip students and experts with advanced topics and upcoming concepts in this area.

After months of intensive research and writing, this book is the end result of all who devoted their time and efforts in the initiation and progress of this book. It will surely be a source of reference in enhancing the required knowledge of the new developments in the area. During the course of developing this book, certain measures such as accuracy, authenticity and research focused analytical studies were given preference in order to produce a comprehensive book in the area of study.

This book would not have been possible without the efforts of the authors and the publisher. I extend my sincere thanks to them. Secondly, I express my gratitude to my family and well-wishers. And most importantly, I thank my students for constantly expressing their willingness and curiosity in enhancing their knowledge in the field, which encourages me to take up further research projects for the advancement of the area.

Editor

1

Simplified Multi-Block Constitutive Model Predicting the Seismic Displacement of Saturated Sands along Slip Surfaces with Strain Softening

Constantine A. Stamatopoulos

1. Introduction

Slopes consisting of saturated sands have recently moved down-slope tens or hundreds of meters under the action of earthquakes [1-3]. This large slide movement is usually associated with the generation of large excess pore pressures, as a result of grain crushing [3]. As these landslides have caused much destruction and fatalities, there is a need to propose easy-to-use and cost-effective methods predicting the triggering and movement of such slides.

The sliding-block model [4] is frequently used to simulate movement of slides triggered by earthquakes [5]. When ground displacement is large, this model may be inaccurate because of (a) reduction of shear resistance along the slip surface and (b) rotation of the sliding mass towards a more stable configuration [6]. It has been modeled in a cost-effective manner by an iterative procedure using the Jambu stability method by Deng et al [7], as well as by the multi-block model by Stamatopoulos et al. [8]. The multi-block model, described below, has the advantage of ensuring displacement compatibility during motion and will be applied in the present work.

Regarding effect (a) above, recently ring shear devices where sandy samples can be sheared under undained conditions have been developed and applied to study the response of saturated sands along slip surfaces [2], [3], [9-15]. Constitutive equations modeling this soil response coupled with the multi-block sliding system model are needed in order to simulate the triggering of the slides and predict accurately the slide displacement. In the general case, constitutive models must be formulated in terms of effective stress in order to predict not only the shear stress, but also the generation of excess pore pressure along slip surfaces. Gerolymos and Gazetas [16] proposed an effective stress model predicting the displacement of saturated sands along slip surfaces based on grain crushing theory which requires 9 model parameters.

In addition, Stamatopoulos and Korai [17] proposed and validated with a number of ring shear tests a constitutive model in terms of effective stresses simulating the change of resistance along slip surfaces with shear displacement for sands either under undrained or drained conditions. The model requires 12 model parameters evaluation, which require the availability or performance of ring shear tests at different densities and drainage conditions.

On the other hand, in sliding-block models only the shear resistance versus shear displacement soil response affects the solution. This response depends on the drainage conditions and may alter as a result of dissipation of excess pore pressure in saturated sand. Yet, under earthquake-induced slides, triggering and slide movement are so rapid that dissipation of excess pore pressure does not occur and saturated sands behave in an undrained manner [1-3], [16]. It is inferred that for predicting the triggering and the slide movement of such slides, only the shear resistance versus shear displacement soil response under undrained conditions may be predicted.

The purpose of the present chapter is to present a cost-effective, but accurate, method predicting the triggering and displacement of slides consisting of saturated sands during earthquakes. For this purpose, first the shear stress -displacement response of saturated sands along slip surfaces is described and a model predicting this response with the minimum number of parameters is proposed, calibrated and validated. Then, this constitutive model is coupled with the multi-block sliding system model and the steps needed to predict earthquake-induced slide triggering and large displacement along slip surfaces of saturated sand using this improved model are specified in detail. Finally, the new improved method is validated by application at the well-documented Higashi Takezawa earthquake-induced slide.

2. Soil response

For the simulation of the response of sands along slip surfaces the ring shear device has the advantage that it is the only device where, similarly to landslides, shear displacement can be very large, larger than centimeters, or even meters. The sample in the ring-shear box is doughnut shaped. Undrained response is maintained by pressing a necessary contact pressure in order to keep the specimen volume constant. Additional details on the ring shear apparatus are given by Sassa et al. [18]. Ring-shear tests under both monotonic and cyclic loading have been performed and illustrated that the virgin shear stress-displacement response of saturated sands is almost identical under monotonic and cyclic loading [13]. This investigation also illustrated that the unloading-reloading stiffness is much larger than that due to virgin loading, especially when shear displacement that has already accumulated is considerable. Typical results are given in Fig. 1.

Table 1 gives tests found in the literature where saturated sands are sheared under undrained conditions in the ring shear device, the soils used in these tests, the initial density and stress conditions and the relevant references. Table 2 gives their particle diameter D50 and D10. Fig 2 gives the measured shear stress-displacement response of these tests. In addition, table 1 and Fig. 3 gives two tests performed in a simple shear device that represent adequately the undrained response of the saturated sand along the slip surface of the Higashi Takezawa landslide [1]. It can be observed that in all tests the shear stress first increases at a decreasing rate with shear displacement and reaches its peak value in a few millimeters or centimeters. Then, it decreases at a decreasing with shear displacement rate and reaches its residual value in many centimeters or meters.

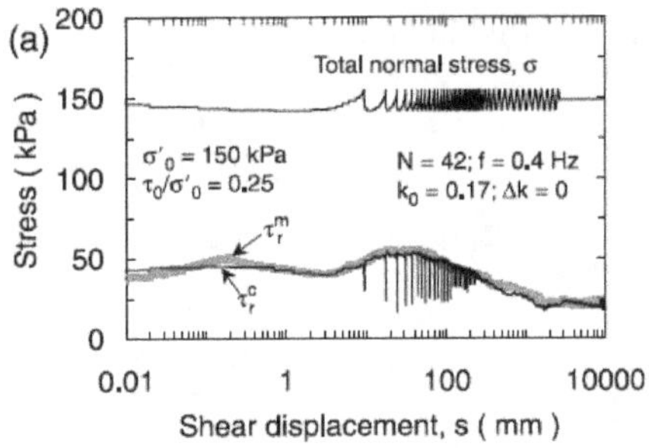

Figure 1. Typical measured effect of cyclic loading on shear stress-displacement relationship. (Trandafir and Sassa [13]).

Test no	Ref.	Sand name	Dr%	σ'o (kPa)	τo (kPa)
1	[9]	S8	62	196	56
2	[9]	S8	63	196	0
3	[9]	S8	63	196	4
4	[10]	S8	63	200	0
5	[9]	S8	65	196	56
6	[9]	S8	68	196	56
7	[9]	S8	68	196	0
8	[9]	S8	69	196	40
9	[9]	S8	69	196	75
10	[9]	S8	69	196	110
11	[9]	S8	95	196	0
12	[11]	ING	29	202	0
13	[11]	ING	29	262	0
14	[11]	ING	32	290	0
15	[12]	ING	44	196	0
16	[12]	ING	44	280	9
17	[12]	ING	44	374	9
18	[11]	WG	30	196	0
19	[11]	WG	30	250	0
20	**[11]**	**WG**	**30**	**290**	**0**
21	**[12]**	**WG**	**44**	**203**	**0**
22	**[12]**	**WG**	**44**	**235**	**0**
23	**[12]**	**WG**	**44**	**290**	**0**
24	**[12]**	**WG**	**44**	**366**	**0**
25	**[1]**	**Higashi**		**50**	**35**
26	**[1]**	**Higashi**		**99**	**40**

Table 1. Constant-volume ring shear tests in saturated sands found in the literature. Relevant reference (Ref.), sand name and initial conditions are given.

	WG	ING	S8
D50 (mm)	0.23	0.11	0.06
D10 (mm)	0.04	0.03	0.015

Table 2. D50 and D10 values of the sands reported in table 1.

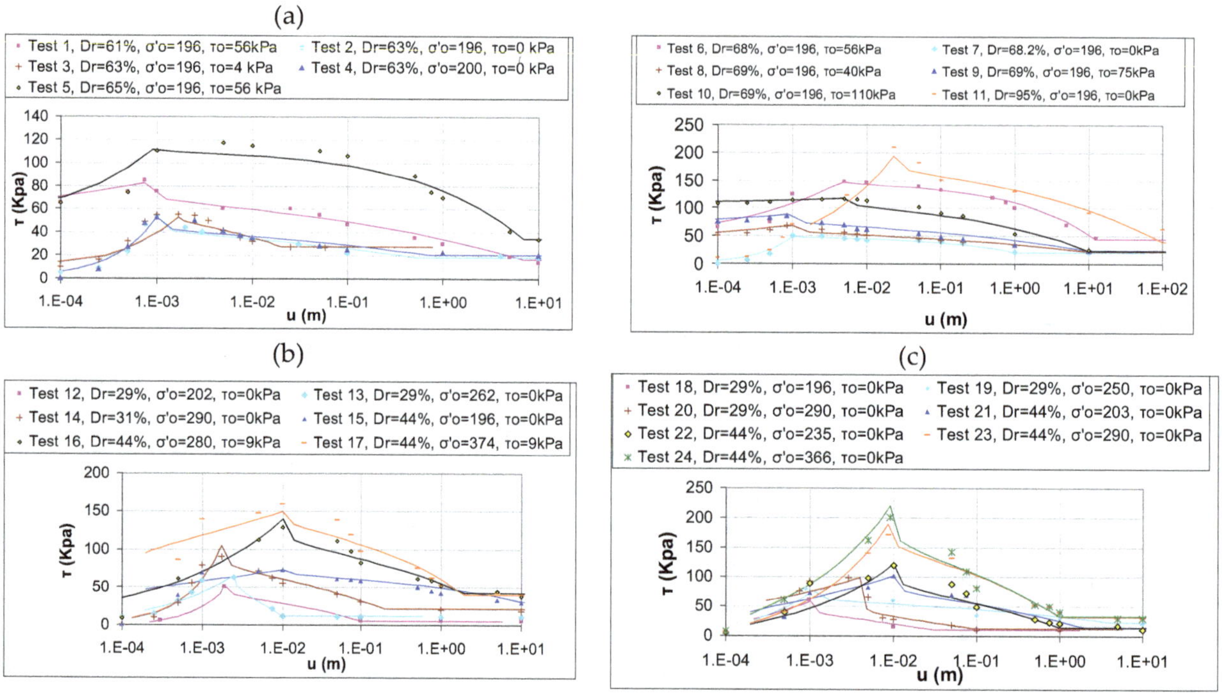

Figure 2. Shear stress-displacement response of the constant-volume ring shear tests of table 1. Tests on sands (a) S8, (b) ING, (c) WG. Model predictions are also given.

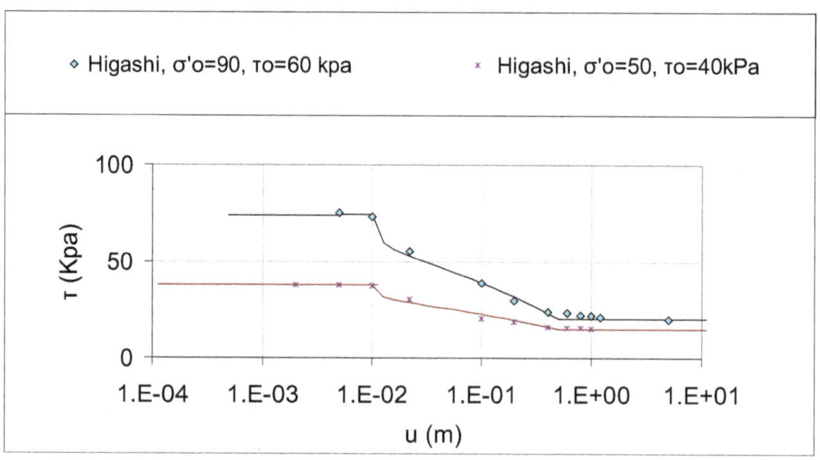

Figure 3. Shear stress-displacement response of the constant-volume ring shear tests of table 1. Tests on the Higashi Takezawa landslide. Model predictions are also given.

3. Proposed model

The following simplified equations are used to simulate shear stress-displacement response of sands under undrained conditions:

$$\begin{aligned}
&\tau = \sigma'_0 R \text{ where} && \text{(a)}\\
&\text{For } 0<u<u_1 \quad R=R_{res}\frac{u^{a_1}}{r\,u_1^{a_1}} && \text{(b)}\\
&\text{For } u_1>u>u_2 \quad R_{res}\{\frac{1}{r}-(\frac{1}{r}-1)\frac{(u-u_1)^{a_2}}{(u_2-u_1)^{a_2}}\} && \text{(c)}\\
&\text{For } u>u_2 \quad R=R_{res} && \text{(d)}
\end{aligned} \qquad (1)$$

where σ'_o is the applied effective normal stress and u is the shear displacement. The model has 6 parameters: R_{res}, r, u1, u2, a1, a2. As illustrated in Fig. 4, R_{res} is the stress ratio (τ/σ'_o) of the material at the residual state. The parameter r equals to (R_{res}/R_{max}). Thus, it varies from 0 to 1. The parameters (a) u_1 and (b) u_2 give the shear displacement corresponding to (a) the peak shear stress and (b) the minimum shear displacement corresponding to the residual shear stress. The parameters (i) a1 and (ii) a2 determine the rate of change of the shear stress with shear displacement for (i) u1<u and (ii) u1<u<u2 respectively.

In the proposed constitutive model plastic shear displacements are assumed to accumulate when the stress ratio (τ/σ') increases, while the elastic shear displacement is ignored. This is consistent with the response depicted in Fig. 1.

An excel worksheet was programmed to simulate the undrained response of sands, as described by equations (1). A modification is needed for tests with initial shear stress (τ_o). In this case the shear displacement is u' and equals:

$$u' = u - u_0 = u - u_1\left[\frac{\tau_0}{\sigma'_0 R_m}\right]^{\frac{1}{a_1}} \quad u'>0 \qquad (2)$$

where u is the displacement predicted by eq. (1).

The model parameters of the tests of Table 1 were estimated using the procedure described above. They are given in Table 3. In addition, Fig. 2 gives the computed shear stress-displacement curves. Table 3 gives the standard deviation of the ratio of predicted by measured values of all points defining the shear stress-displacement curves of all tests. It can be observed that in all tests the standard deviation is less than 0.3, while the average value is 0.12. In addition, from Fig 2 it can be observed that adequate prediction of undrained response is achieved in all tests. Furthermore, from table 3 it can be observed that the values of the model parameters used do not vary considerably from case to case and are in general in a rational range: the parameters a1 and a2 do not take values greater than 1 and less than 0 and the parameter Rm generally increases with Dr. All the above verify the proposed model.

Advantage of the proposed model is simplicity. However, it has the disadvantage of generality. Inspection of table 3, illustrates that the model parameters depend on the relative density, confining stress and initial shear stress. Thus, when applying this simplified model, tests with similar relative density, confining stress and initial shear stress as that existing in-situ should be used.

Test no	Rm	r	u1 (m)	u2 (m)	a1	a2	N	SDev (Pred/Meas)
1	0.42	0.20	0.001	7.00	0.08	0.16	12	0.07
2	0.27	0.35	0.001	0.50	1.00	0.15	11	0.17
3	0.27	0.52	0.002	0.02	0.57	0.48	12	0.15
4	0.27	0.39	0.001	0.70	1.00	0.15	10	0.12
5	0.57	0.31	0.001	7.00	0.30	0.40	12	0.18
6	0.75	0.30	0.005	12.50	0.20	0.40	12	0.13
7	0.25	0.42	0.001	1.00	1.00	0.38	10	0.12
8	0.35	0.32	0.001	10.00	0.10	0.14	12	0.07
9	0.45	0.26	0.001	10.00	0.08	0.15	12	0.10
10	0.60	0.20	0.005	10.00	0.02	0.24	12	0.07
11	0.99	0.20	0.023	100.00	0.40	0.20	9	0.15
12	0.24	0.10	0.002	0.10	1.00	0.30	5	0.01
13	0.25	0.17	0.003	0.01	0.50	0.40	11	0.17
14	0.36	0.23	0.002	0.50	1.00	0.15	10	0.12
15	0.37	0.42	0.010	10.00	0.10	0.25	1	0.14
16	0.50	0.30	0.010	2.00	0.30	0.20	10	0.18
17	0.40	0.26	0.010	2.00	0.10	0.30	11	0.14
18	0.30	0.17	0.001	0.03	0.35	0.20	6	0.14
19	0.30	0.30	0.001	4.00	0.50	0.15	4	0.16
20	0.34	0.12	0.004	0.10	0.20	0.10	9	0.18
21	0.50	0.13	0.010	2.00	0.20	0.20	10	0.18
22	0.52	0.12	0.010	1.00	0.40	0.20	11	0.17
23	0.65	0.17	0.009	1.00	0.50	0.25	10	0.14
24	0.60	0.15	0.009	1.00	0.40	0.20	9	0.15
25	1.10	0.28	0.020	10.00	0.40	0.25	5	0.05
26	0.60	0.28	0.010	5.00	0.05	0.20	6	0.04
						All Mean		0.12
						All SDev		0.05

Table 3. The parameters of equation (1) that fit the tests of table 1. Statistical analysis of the accuracy of the predictions is also given, where N represents the number of points defining the shear stress-displacement curve of each test.

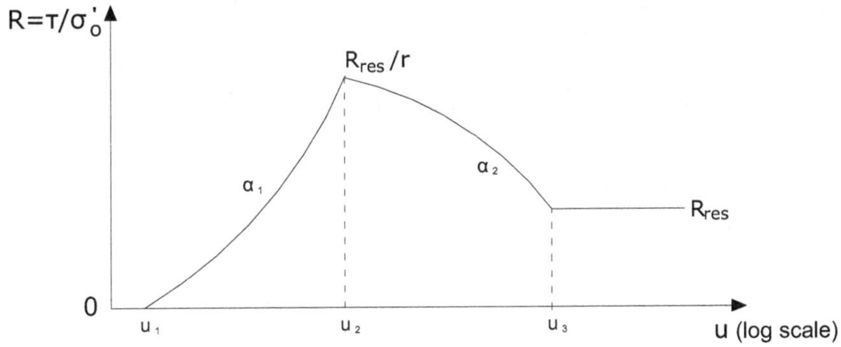

Figure 4. The proposed model and its model parameters.

4. The Multi-Block Sliding System Model

4.1. Introduction

A slip surface which consists of linear segments is considered (Fig. 5a). In order for the mass above the slip surface to move, interfaces inside the sliding mass must be formed at the nodes between the linear segments [20]. In this manner the mass is divided into n blocks. Soil is assumed to behave as a rigid-perfectly plastic material with a Mohr-Coulomb failure criterion at both the slip surface and the interfaces. The forces that are exerted in block "i" are given in Fig. 5a.

When the slide moves, two options exist regarding the relative movement of blocks: (a) no separation and (b) separation. When blocks are not separated, the velocity must be continuous at the interface. This rule predicts that the relative displacement of the n blocks is related to each other as:

$$\frac{du_i}{du_{i+1}} = \frac{\cos(\delta_i + \beta_{i+1})}{\cos(\delta_i + \beta_i)} \tag{3}$$

where u is the displacement moved by the sliding mass along the linear segment "i" of the trajectory, d refers to increment, the subscripts i and i+1 refer to trajectory segments i and i+1 counting uphill and β_i and $(90-\delta_i)$ are the inclinations of the trajectory segment and interface i respectively.

Fig. 5a illustrates the forces exerted in each block i. In the case that separation of blocks does not occur and a horizontal acceleration is applied, as indicated by Stamatopoulos et al [8], the governing equation of motion is

$$\ddot{u}(t)\cdot\sum_{i=1}^{n}\left(m_i q_i \cos(\phi_i)\cdot\prod_{j=i}^{n-1}\frac{dd_{j+1}}{ff_j}\right)=\sum_{i=1}^{n-1}\left(\left[-bc_i\cos(\phi b_i)+Ub_i\sin(\phi b_i)\right]\cdot\frac{ss_i}{ff_i}\cdot\prod_{j=i}^{n-1}\frac{dd_{j+1}}{ff_j}\right)+$$

$$+\sum_{i=1}^{n}\left(\left\{xx_i\left(m_i g k(t)+H_i\right)-v_i\left[m_i g+Q_i\right]-c_i l_i\cos(\phi_i)+U_i\sin(\phi_i)\right\}\cdot\prod_{j=i}^{n-1}\frac{dd_{j+1}}{ff_j}\right) \quad \text{(a)}$$

where

$$ss_i=\sin\left(\beta_{i+1}-\beta_i+\varphi_i+\varphi b_{i+1}\right) \quad \text{(b)}$$

$$\mathrm{ff}_i=\cos\left(\varphi_i+\varphi \mathrm{b}_{i+1}-\beta_i-\delta_i\right) \quad (4)$$

$$\mathrm{xx}_i=\cos\left(\varphi_i-\beta\right) \quad \text{(c)}$$

$$\mathrm{v}_i=\sin\left(\varphi_i-\beta_i\right) \quad \text{(d)}$$

$$\mathrm{dd}_i=\cos\left(\varphi_i+\varphi \mathrm{b}_i-\beta_i-\delta_{i-1}\right) \quad \text{(e)}$$

$$\mathrm{q}_i=\tilde{\mathrm{O}}_{j=i}^{n-1}\frac{\delta\mathrm{os}\left(\beta_{i+1}+{}_{i}\right)}{\delta\mathrm{os}\left(\beta_i+{}_{i}\right)} \quad \text{(f)}$$

where u is the relative shear displacement of the uppermost block (block n), down-hill displacements being considered positive, g is the acceleration of gravity, Q_i is the vertical external load on block i, H_i is the horizontal external load on block i, m_i is the mass of block i, U_i is the pore water force acting on block i and Ub_i is the pore water force along the interface between blocks i and i+1, φ_i and c_i is the angle of friction and cohesion on the slip surface at block i along the slip surface and φb_i and cb_i is the angle of friction and cohesion on the interslice surface between blocks i and i+1. It should be noted that all the above masses and forces are taken per unit length, normal to the paper in Fig. 5a.

Equation (4) can be written in the form

$$\ddot{u}_1 = Z_1\, g\,(k(t)-k_c)\ \text{where} \quad \text{(a)}$$

$$Z_1=\frac{\sum_{i=1}^{n}\left(-xx_i m_i\cdot\prod_{j=i}^{n-1}\frac{dd_{j+1}}{ff_j}\right)}{\sum_{i=1}^{n}\left(m_i q_i\cos(\phi_i)\cdot\prod_{j=i}^{n-1}\frac{dd_{j+1}}{ff_j}\right)} \quad \text{(b)}$$

$$k_c=\frac{\sum_{i=1}^{n-1}\left(\left[cb_i b_i\cos(\phi b_i)-Ub_i\sin(b\phi_i)\right]\cdot\frac{ss_i}{ff_i}\cdot\prod_{j=i+1}^{n-1}\frac{dd_{j+1}}{ff_j}\right)+}{g\cdot\sum_{i=1}^{n}\left(-xx_i m_i\cdot\prod_{j=i}^{n-1}\frac{dd_{j+1}}{ff_j}\right)} \quad (5)$$

$$+\frac{\sum_{i=1}^{n}\left(\left\{-xx_i(H_i)+v_i\left[W+Q_i\right]+c_i l_i\cos(\phi_i)-U_i\sin(\phi_i)\right\}\cdot\prod_{j=i}^{n-1}\frac{dd_{j+1}}{ff_j}\right)}{g\cdot\sum_{i=1}^{n}\left(-xx_i m_i\cdot\prod_{j=i}^{n-1}\frac{dd_{j+1}}{ff_j}\right)} \quad \text{(c)}$$

where k_c is the critical acceleration factor, defined as the limit horizontal acceleration needed to initiate motion (a_c) normalized by the acceleration of gravity (g).

For large displacement, the masses and lengths of each block in equation (5) must be updated in terms of the distance moved. Assuming that the displacement u is less than the initial length of l_n, the change of lengths l_i in each displacement increment Δu equals

$$\begin{aligned} &\Delta l_i = \Delta u \cdot q_1 && \text{(a)} \\ &\Delta l_n = -\Delta u && \text{(b)} \\ &\Delta l_2 = \Delta l_3 = \ldots.. = \Delta l_{n-1} = 0 && \text{(c)} \end{aligned} \tag{6}$$

The incremental change in the interslice lengths b_i is

$$\Delta b_i = \frac{\sin\theta_i}{\cos(\theta_i + \beta_i + \delta_i)} \cdot q_i \Delta u \tag{7}$$

where the angle θ_i is given in Fig. 5b. Fig. 5c illustrates the deformation that these rules predict. In addition, the incremental change in the mass is

$$\begin{aligned} \Delta m_i = \rho_{i+1}\{ & b_{(i+1)-c} \cdot \cos(\beta_{i+1} + \delta_{i+1}) \cdot q_i \Delta u + \frac{0.5 \cdot \cos(\beta_{i+1} + \delta_{i+1}) \cdot \sin\theta_{i+1}}{\cos(\theta_{i+1} + \beta_{i+1} + \delta_{i+1})} \cdot (q_i \Delta u)^2 \\ & - b_{i-c} \cdot \cos(\beta_i + \delta_i) \cdot q_i \Delta u - \frac{0.5 \cdot \cos(\beta_i + \delta_i) \cdot \sin\theta_i}{\cos(\theta_i + \beta_i + \delta_i)} \cdot (q_i \Delta u)^2 \} \end{aligned} \tag{8}$$

where b_{i-c} and $b_{(i+1)-c}$ correspond to the values of b_i and $b_{(i+1)}$ at the previous increment and ϱ is the total unit weight of the soil.

Separation of blocks occurs when an interslice force, N_i, is negative [21]. Fig. 6c illustrates a typical case where this occurs: when the angle β_m at the trajectory is larger than the angle β_{m+1}. In this case, the soil mass of the block along the "m+1" segment of the trajectory cannot maintain contact with the rest of the sliding material and is detached from the system. The detached mass is no longer considered in the solution.

4.2. Multi-block model with constitutive equations

The multi-block sliding system can be coupled with the constitutive model described above by assuming zero cohesion and varying only the friction angles, φ_i, of equation (5) as

$$\varphi_i = \arctan\left(\frac{\tau_i}{\sigma'_{o-i}}\right) \tag{9}$$

In equation (9), τ_i and σ'_{o-i} are the shear stress and the initial (prior to slide movement) effective normal stress at the base of block i.

Application of eq. (9) with the multi-block numerical code first requires to estimate the shear initial stresses τ_i, or equivalently, the initial friction angles φ_i. This is performed by iterations, by assuming all φ_i are equal and by increasing them incrementally until critical equilibrium is achieved in the initial slide configuration. Then, at each increment, for each block, τ_i is updated from equations (1) in terms of the incremental shear displacement. It should be noted that the values of τ, u and σ'_o of equations (1) correspond to τ_i, u_i and σ'_{o-i} in equation (9).

4.3. Computer program

A computer program which solves the equations described above has been developed by the author. The input geometry is specified as the nodes of the linear segments defining the trajectory, ground and water table surfaces. Different soil properties may be specified at each segment of the slip surface and at the interfaces. Along the slip surface the Mohr-Coulomb Model or the constitutive model may be applied. At the interfaces the Mohr-Coulomb Model is applied. Output of the program includes the final slide geometry and acceleration velocity and displacement of nodes of the sliding mass versus time.

Pore pressures at the mid-points of the linear segments of the linear segments of the slip and the interfaces prior to the application of earthquake loading are estimated from the water table surface according to the general equation:

$$P_i = h_{W-i} \cos^2(\theta w_i) \tag{10}$$

where θw_i is the inclination and h_i is the height of the water table surface above the mid-point of slip segment "i". The program includes graphical representations of the input and final geometries of the slope. More details of the multi-block method and the associated numerical code are given by Stamatopoulos et al [8].

4.4. Application of the model along (pre-defined) slip surfaces and very large displacement under earthquake loading

In the case that the slip surface is not pre-existing, generally, application of the sliding-block model first requires the prediction of the location of the slip surface by stability analysis. However, this determination, and the ability of stability methods to estimate the location of this slip surface is beyond the purpose of the present work.

In order to apply the improved multi-block model along pre-defined slip surfaces under earthquake loading first the slip surface, ground surface and water table surface are simulated as a series of linear segments. If the inclinations of the interfaces is not predifined according to existing faults, as proposed by Sarma [20] they are obtained based on the condition of the minimum critical acceleration value, by iterations. Along the slip surface the Mohr-Coulomb model the value is used with soil strength corresponding to large displacement, as it is the most representative of the soil strength during occurrence of the landslide. At the interfaces, peak values of strength are used. The reason is that as the internal interfaces are fixed in space, they are continuously reforming with new material and thus the strength cannot be at residual [23]. In addition, at large deformations the method described above to estimate the interslice angles of the sliding mass according to the condition of minimum critical acceleration at the initial slide configuration may not be adequate. The reason is that some segments of the trajectory do not have mass at the initial configuration and thus their interface angles cannot be defined. This can be resolved by applying the criterion of minimum acceleration not only at the initial, but also at the final slide configuration and taking the average values at the common interfaces of the two interfaces. The final slide configuration can be obtained by applying the multi-block model assuming that the interface angles not defined at the initial slide configuration equal zero.

The slide triggering is investigated and the potential slide deformation is estimated using the multi-block model. For the representative seismic motion, the proposed constitutive model is used along the slip surface, while at the interfaces the Mohr-Coulomb model with peak values of strength is used.

5. Application at the Higashi Takezawa landslide

5.1. The slide

Fig. 6 gives the cross-section of the Higashi Takezawa landslide [1]. According to Deng et al. [1] a representative acceleration record for the slide is that reported by the Japan Meteorological Agency (JMA) in 2004. It is given in Fig. 6. Triaxial and shear laboratory tests of material along the slip surface of the slide have been performed by Deng et al., [1] and are given in Fig. 6.

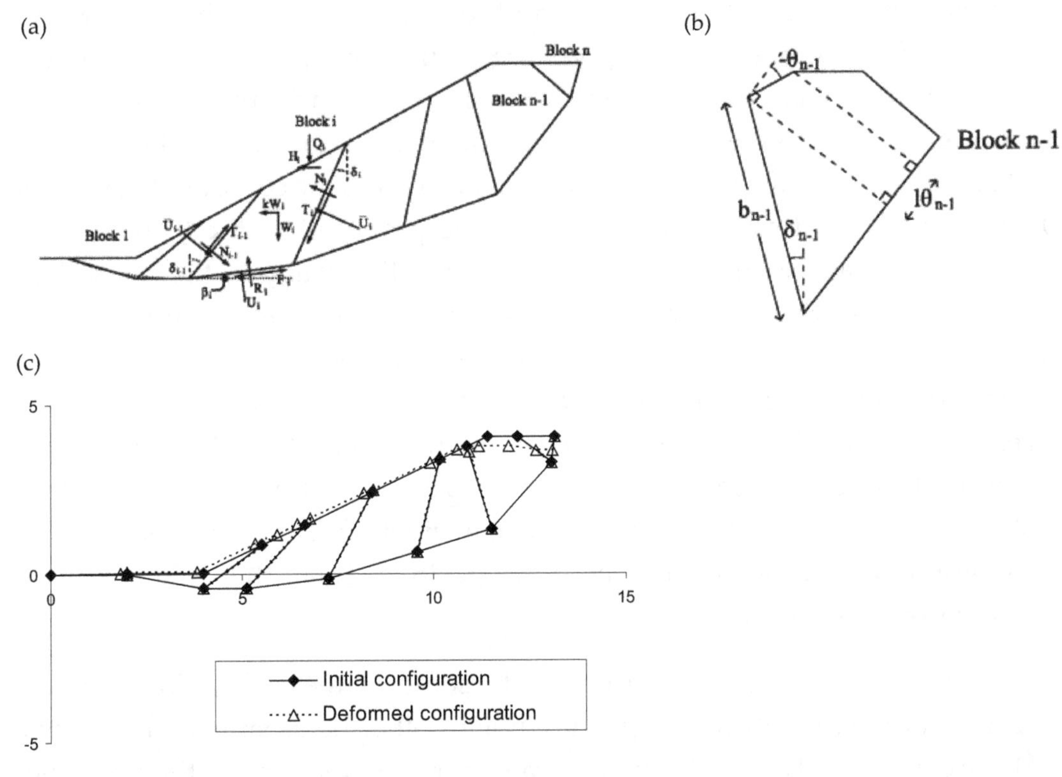

Figure 5. (a) The multi-block stability method proposed by Sarma (1979). (b) Definition of the angle θ_i of the ground surface of block i. The angle θ_i changes when $u_i > u\theta_i$ (c) Deformation assumed in the multi-block model for a case of two blocks. The x-axis gives the horizontal distance, while the y-axis gives the elevation.

5.2. Model predictions

First the model parameters of the constitutive model are obtained by the prediction of the shear test results. Table 4 gives the model parameters and Fig. 7 compares model predictions of the constitutive model with the measured response. Very good agreement is observed. It can be observed that even though all model parameters do not depend on the applied normal stress, predictions are adequate.

From Fig. 6 it can be observed that all of the slip surface is below the water table line. In addition, according to Deng et al. [1] the material at the slip surface is approximately uniform. Thus, the model parameters of Table 4 are used to simulate the soil response at the slip surface. At the interfaces, the peak soil strength must be used, which, according to the results of the triaxial tests, equals $\varphi'=36^{o}$.

Typically, the criterion of minimum critical acceleration factor at the initial and final configurations is applied to estimate the interface angles of the sliding mass. The Mohr Coulomb model was applied both along the slip surface and at the interfaces. Along the slip surface the residual soil strength was applied, while at the interface the peak soil strength. Fig. 8 gives the critical acceleration in terms of the interslice angles at the (a) initial and (b) final slide configurations. Table 5 gives the obtained interslice angles according to the criterion of minimum acceleration factor and the results of Fig. 8.

Once the interface angles are obtained, the multi-block model with the constitutive model along the slip surface is applied. Fig. 9a gives the input geometry used to simulate the slide with the multi-block model. As the seismic motion of Fig. 6d is applied, Fig. 10 gives the computed equivalent friction angle of block 2 (equation (9) and the slide acceleration (a), velocity (V) and distance moved (u). In addition, Fig. 9b gives the computed versus measured final slide configuration. It can be observed that the proposed method predicted both the triggereing of the slide and with good accuracy the final slide configuration. From Fig. 9 it can be observed, as the earthquake is applied, some shear displacement accumulates. This causes the friction angle at the base of the blocks to increase. Once the peak friction angle is reached, due to material softening, the friction angle decreases, to its residual value. At this point, the critical acceleration of the sliding system is negative (this means that the slide is unstable) and the slide velocity starts to increase and displacement to accumulate rapidly. As the slide moves, the mass slides at a progressively smaller average inclination. The critical acceleration of the sliding mass gradually increases and it becomes positive. Then, the slide velocity starts to decrease, and becomes zero at t=51s.

Finally, parametric analyses were performed. In some of these the Mohr-Coulomb with residual or maximum strength was applied, while in other no seismic motion was applied. Their results are given in Table 6. They illustrate that even though for the prediction of the triggering of the slide the applied motion and the constitutive model are important, for the prediction of the final slide configuration, use of the Mohr-coulomb model with the residual soil strength produces relatively accurate results, even when the seismic motion is not applied.

Rm	r	u1 (m)	u2 (m)	a1	a2
0.8	0.3	0.01	0.5	0.3	0.25

Table 4. The Higashi Takezawa landslide. Model parameters

i=	Initial config.	Final config.	Mean value
1	-	30°	30°
2	-20°	10°	-5°
3	-10°	-	-10°

Table 5. The Higashi Takezawa landslide. Obtained interslice angles (δi) according to the criterion of minimum acceleration and the result of Fig. 8.

Seismic Motion	Soil model	u (m)
Yes	Constitutive	376
Yes	Mohr-Coulomb with residual strength	414
Yes	Mohr-Coulomb with maximum strength	0.25
No	Mohr-Coulomb with residual strength	427
No	Mohr-Coulomb with maximum strength	0

Table 6. Parametric analyses. Slide displacement in terms of the soil model and seismic motion applied.

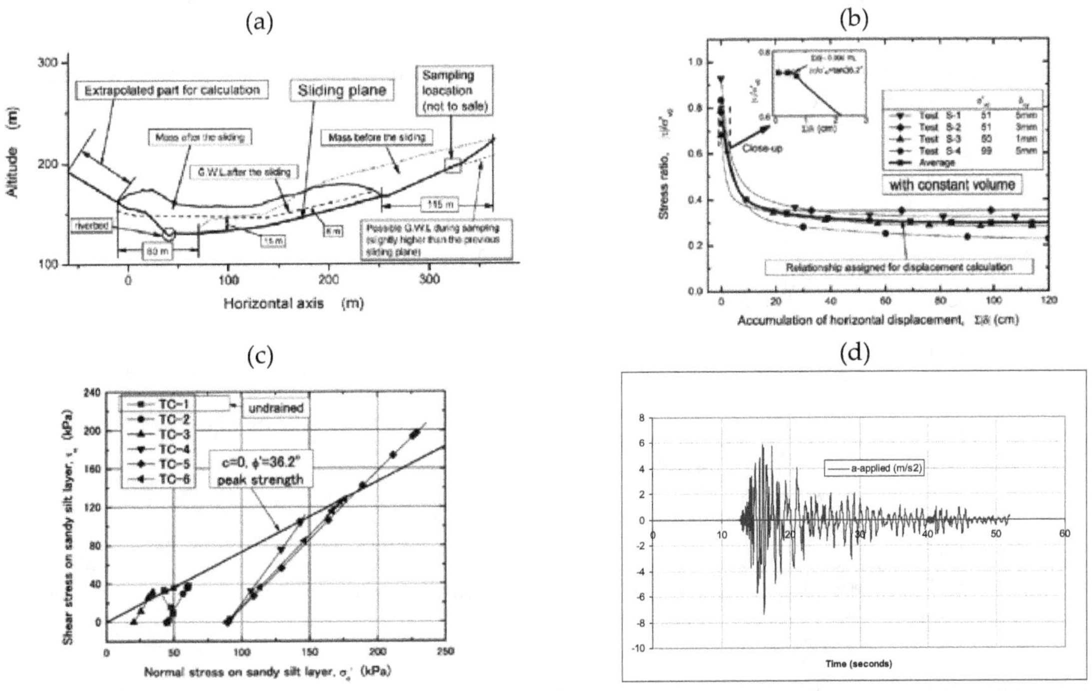

Figure 6. The Higashi Takezawa landslide. (a).Cross-section of slide (Deng et al. [1], (b). Simple shear test results (Deng et al. [1], (c) Triaxial test results (Deng et al. [1], (d) representative applied acceleration in terms of time (Deng et al. [1]

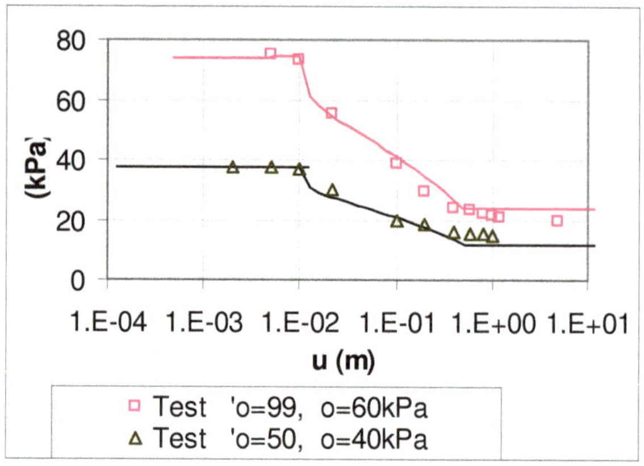

Figure 7. The Higashi Takezawa landslide. Computed versus measured response in terms of initial consolidation stress for the unique model parameters of table 4.

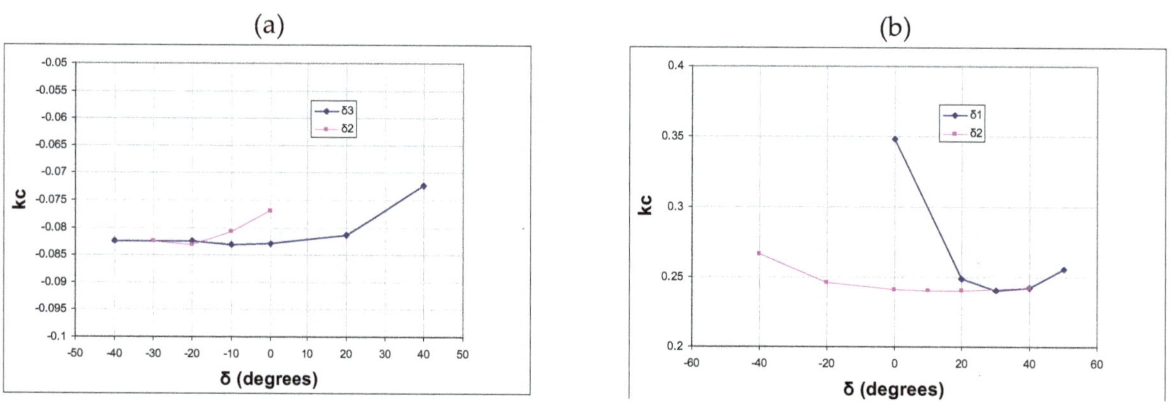

Figure 8. The Higashi Takezawa landslide. Critical acceleration in terms of the interslice angles. (a) Initial slide configuration, (b) Final slide configuration.

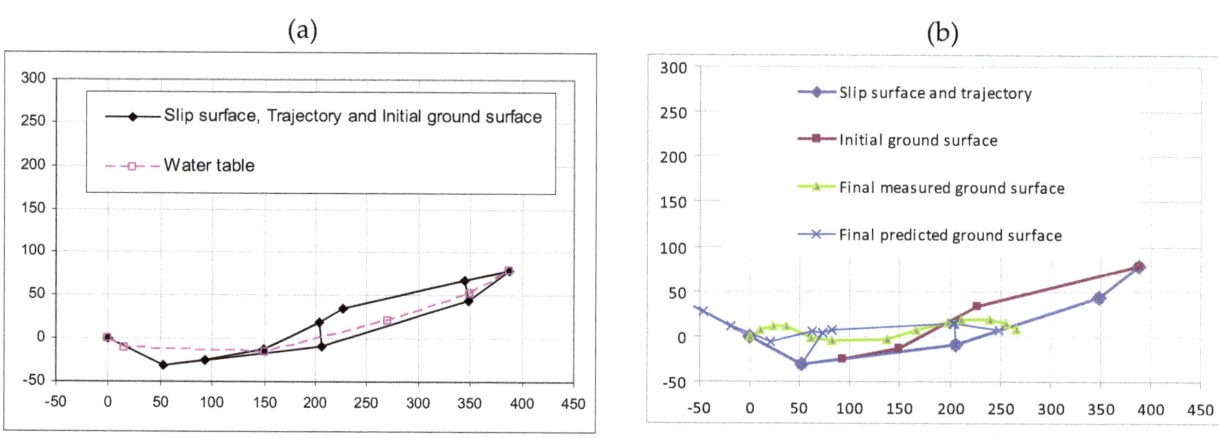

Figure 9. The Higashi Takezawa landslide. (a). Input used to simulate the slide with the multi-block model, (b). Computed versus measured final slide configuration.

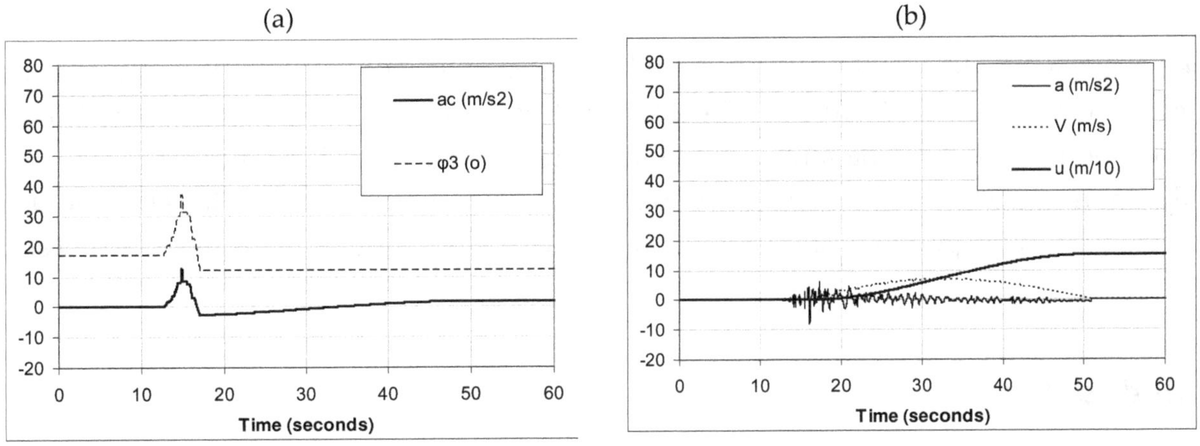

Figure 10. (a). Computed equivalent friction angle of block 3 and critical acceleration of the slide (ac) and (b) slide acceleration (a), velocity (v), distance moved (u), all in terms of time

6. Conclusions

Slopes consisting of saturated sands have moved down-slope tens or hundreds of meters under the action of earthquakes recently. The chapter presents a simplified method predicting the triggering and displacement of such earthquake-induced slides of saturated sands.

For this purpose, a simplified constitutive model simulating soil response of saturated sands along slip surfaces is proposed. Comparison of the model predictions with results of ring shear tests illustrated that the model predicts with good accuracy the shear stress-displacement response of saturated sands. Advantage of the proposed model is simplicity. However, it has the disadvantage of generality. Thus, when applying this simplified model, tests with similar relative density, confining stress and initial shear stress as that existing in-situ should be used.

Then, this constitutive model was coupled with the multi-block sliding system model to predict the triggering and displacement of earthquake-induced slides of saturated sands. The multi-block model considers a general mass sliding on a slip surface which consists of n linear segments. In order for the mass to move, at the nodes between the linear segments, interfaces inside the sliding mass must be formed. Steps needed to apply the method are: (a) define the trajectory, ground and water table surfaces, (b) obtain the model parameters of the soil resistance by the results shear tests, (c) obtain the interface angles by applying the principle of minimum critical acceleration factor and (d) to simulate the triggering and displacement of the slide apply the multi-block model for a representative seismic motion is applied.

The method was applied successfully to predict the triggering, the motion and the final configuration of the well-documented Higashi Takezawa earthquake-induced slide.

Acknowledgements

The work was funded partly by the project "Novel methodologies for the assessment of risk of ground displacement" under ESPA 2007-2013 of Greece, under action: Bilateral S & T Cooperation between China and Greece. Graduate student of the Hellenic Open University Kelly Gouma assisted in the analysis of the case study. Graduate students of the Hellenic Open University P. Sidiropoulos and J. Bakratzas assisted in the collection of the laboratory tests.

Author details

Constantine A. Stamatopoulos[1,2*]

Address all correspondence to: k.stam@saa-geotech.gr

1 Hellenic Open University, Athens, Greece

2 Stamatopoulos and Associates Co, Greece

References

[1] Deng J., Kameya H., Miyashita Y., J., Kuwano R., Koseki J. Study on dip slope failure at Higashi Takezawa induced by 2004 Niigata-Ken Chuetsu earthquake. Soils and. Foundations Vol. 51. 2011. No. 5, 929-943. Oct.

[2] Sun P., Wang F., Yin Y., Wu S.. An experimental study of the mechanism of rapid and long run-out landslides triggered by Wenchuan earthquake. Seismology and geology. 2010, Vol. 32., No. 1, Mar.

[3] Sassa, K., Fukuoka, H., Scarascia-Mugnozza, G., Evans, S. Earthquake-induced landslides: Distribution, motion and mechanisms. Special Issue of Soils and Foundations,Japan Geotechnical Society. 1996. 53-64.

[4] Newmark, N. M. Effect of earthquakes on dams and embankments, Geotechnique, Vol. 15, No. 2, London, England. 1965. June, 139-160.

[5] Ambaseys N. and Menu J. Earthquake induced ground displacements, Earthquake engineering and structural dynamics. 1988. 16, 7, 985-1006.

[6] Stamatopoulos, C. A. Sliding System Predicting Large Permanent Co-Seismic Movements of Slopes. Earthquake Engineering and Structural Dynamics. 1996. Vol. 25, No, 10, 1075-1093.

[7] Deng, J., Tsutsumi, Y, Kameya, H., Koseki, J. A Modified Procedure to Evaluate Earthquake-induced Displacement of Slopes Containing a Weak Layer. Soils and Foundations. 2010, 50, 3; 413-420

[8] Stamatopoulos C. A., Mavromihalis C, Sarma S. Correction for geometry changes during motion of sliding-block seismic displacement, ASCE, Journal of Geotechnical and Geoenvironmental Engineering. 2011, 137 (10), 926-938.

[9] Wang G. and Sassa K. Post-failure mobility of saturated sands in undrained load-controlled ring shear tests. Can. Geotech J.. 2002, 39(4), 821–837.

[10] Wang, G., Sassa K. Fukuoka H. and Tada T. Experimental study on the shearing behavior of saturated silty soils based on ring shear tests. Journal of geotechnical and geoenviromental engineering. 2007, 319-333.

[11] Igwe O., Sassa K., Wang F. The influence of grading on the shear strength of loose sands in stress-controlled ring shear tests. Landslides. 2007, 4, 43–51.

[12] Igwe O., Sassa K., Fukuoka H. The geotechnical properties of sands with varying grading in a stressControlled Ring Shear Tests. The Electronic Journal of Geotechnical Engineering. 2009, 14.

[13] Trandafir, A.C. and Sassa, K. Seismic triggering of catastrophic failures on shear surfaces in saturated cohesionless soils. Canadian Geotechnical Journal. 2005, Vol. 42, No. 1, 229-251.

[14] Okada Y., Sassa K., Fukuoka H. Excess pore pressure and grain crushing of sands by means of undrained and naturally drained ring-shear tests, Engineering Geology 75, 2004, 325–343

[15] Sassa K., Wang F. Progress in Landslide science. Department of prevention research of Kyoto. Chapters 6, 7. 2007, 90-106.

[16] Gerolymos N. and Gazetas G. A model for grain-crushing-induced landslides - Application to Nikawa, Kobe 1995, Soil Dynamics and Earthquake Engineering 27. 2007, 803-817.

[17] Stamatopoulos C., Korai C. Constitutive and multi-block modeling of slides on saturated sands along slip surfaces. 7th Hellenic conference of Geotechnical engineering, Athens, Greece. 2014 (in press)

[18] Sassa K., Fukuoka H.. Excess pore pressure and grain crushing of sands by means of undrained and naturally drained ring-shear tests. Engineering Geology. 2004. pp. 325-343

[19] Sadrekarimi, A. and Olson S. M. Shear Band Formation Observed in Ring Shear Tests on Sandy Soils. Journal of Geotechnical and Geoenvironmental Engineering, ASCE. 2010, Vol. 136, No. 2. 366-375.

[20] Sarma S.K. Stability analysis of embankments and slopes. Journal of Geotechnical Engineering ASCE. 1979, Vol.105, No. 12, 1511-1524.

[21] Sarma S.K. and Chlimitzas G. Co-seismic & post-seismic displacements of slopes, 15th ICSMGE TC4 Satellite Conference on "Lessons Learned from Recent Strong Earthquakes". 2001, 25 August, Istanbul, Turkey

[22] Sarma S. K., Tan D. Determination of critical slip surface in slope analysis, *Geotechnique*, 2006, Vol. 56, 539-550

[23] Ambraseys N., Srbulov M. Earthquake induced displacements of slopes, Soil Dynamics and Earthquake Engineering. 1995, 14, 59-71.

2

A Simulation and Evaluation System Oriented to the Emergency Response Effectiveness of the Abrupt Earthquake Disaster

Yan-Yan Huang

1. Introduction

The untraditional emergency disasters, like the 2008 Sichuan earthquake of China, break out with nothing indication, no time to prepare for. These emergency events always bring us the catastrophe with much casualty, homeless and immeasurable economic losses.

The emergency disasters response problem is an important thing to be researched all over the world. As we all know that we can not change the breaking out of the untraditional emergency event in present technology level, but it never means that we could do nothing, on the contrary, if we have enough response plans or preparedness, the disasters should be reduced to the minimum level. It requires us to prepare many valid emergency response plans by hands. However, how to get such valid emergency response plans?

Considering the emergency catastrophe like earthquake takes place so abruptly, it is difficult to make decisions to get the most suitable preparedness plan from many emergency response plans. In order to improve the emergency response capability for the abrupt earthquake disasters, it is necessary to build a decision-making support system to evaluate the response plans and to select a most suitable preparedness plans.

In fact, it is difficult to test and evaluate the response plans in the real conditions of a strong earthquake disaster. As far as we know, modeling and simulation methods are fit for evaluating the preparedness plans. Adopting the modeling and simulation methods, there are many advantages such as low cost, safety research, time-space easily convert, and so on.

Many emergency researches, authors in reference [1-3], show that the more preparedness for the emergency event, the less risk we suffer from the uncertainty emergency. As we all know that, the emergency plans are oriented to the future event, we have none of the real data and sense. So the modeling and simulation methods are utilized to the decision-make support system.

More researches have discussion about the modeling and simulation method in the emergency response management [4]. At the mean time, Many researcher and experts have noticed the importance of the simulation and evaluation method in the decision-making support system [5,6]. The emergency response needs the decision making support system [7,8]·

In China, there are many learners to carry out their researches on the emergency response management, Professor Wang [9] puts forth the parallel simulation method to research the emergency management. Fan Weicheng [10] points to the public incidents, provides some useful suggestion and response methods for the emergency platform system.

In this chapter, we set the research object orient to the emergency earthquake disasters. Design a decision-making support system framework for the emergency response management, and then mainly research the simulation and evaluation methods and their application in the earthquake disasters.

2. Systematic frameworks for simulating and evaluating the emergency preparedness for earthquake

Judging and analyzing the advantage and disadvantages of an emergency response plan needs a good decision making support system. To build up such a decision making system, a serial of technical methods and theory models are required to support the system. In this chapter, the author sums up the research methods for the emergency response plans, and put forth a whole research route map to design and develop the system. Such a research route map is a set of systematic framework which can simulate and evaluate the response plans for the untraditional emergency disaster like earthquake, seeing the Figure 1.

In Figure 1, there are 5 main parts (subsystems) in the set of systematic framework. These subsystems are built to support the whole decision framework. They are including: (A) emergency response effectiveness concept and the evaluation indices, (B) indices weights acquisition,(C)emergency response simulation theory based on OODA-DEVS, (D) simulation system and simulation data acquirement,(E) the integrated evaluation process and decision making support system.

The next research is to describe the related subsystems in the Figure 1. The following research will take emphasis on the main techniques of the simulation and evaluation in emergency response plans.

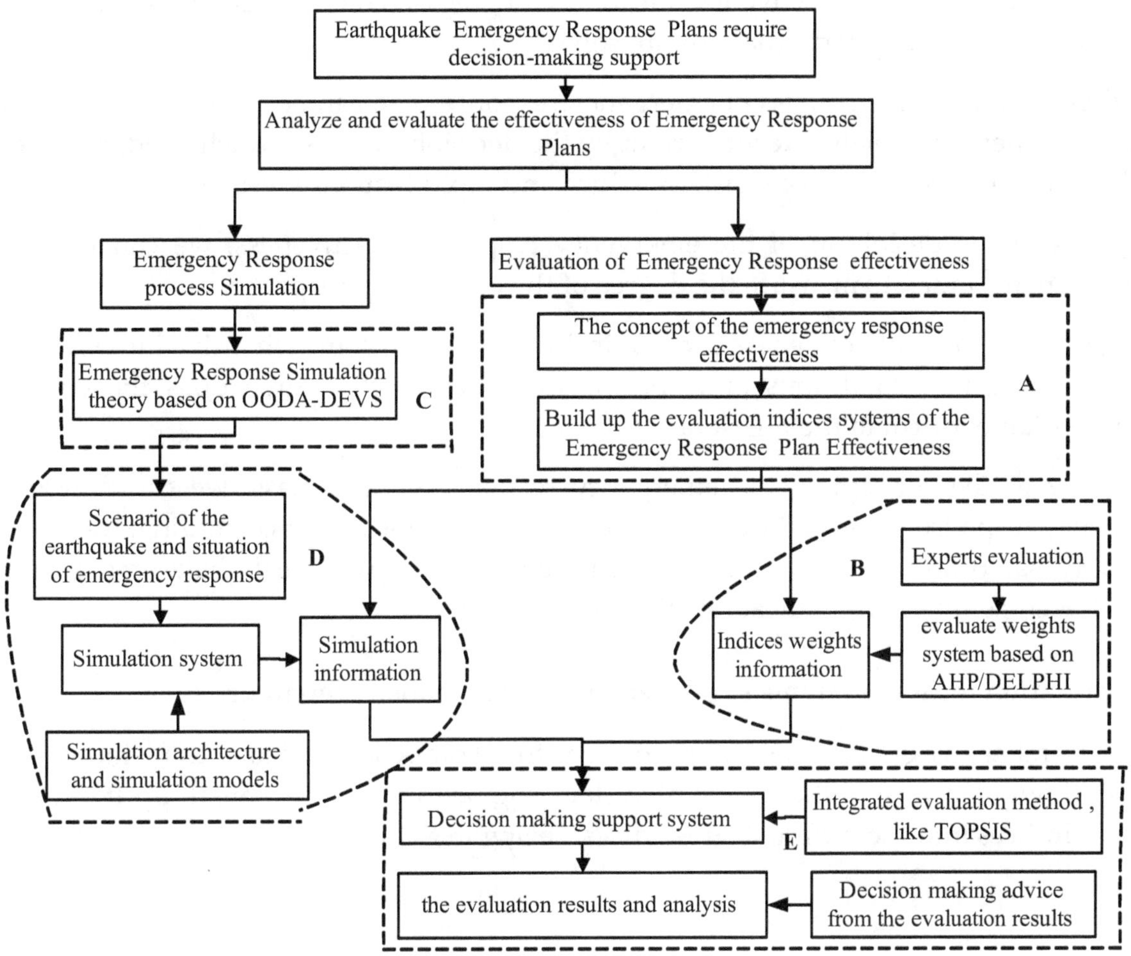

Figure 1. Systematic framework to simulate and evaluate the abrupt earthquake disaster

3. Evaluation indices orient to the emergency response of the earthquake disaster

In order to select a right plan from many emergency response plans, it is necessary to make a good decision to evaluate the emergency response plans. And therefore, build up a set of evaluation indices system is very important.

3.1. Measurements on the effectiveness for emergency response

To evaluate the emergency response plans, we need design a comprehensive evaluation index. Such a complex index is needed to measure what is the degree of an emergency response plan can perform.

The operational effectiveness is widely used in military evaluation. As we know that, a war is very similar to the fight for the earthquake disaster. And thus, we need to give a definition for the effectiveness. In fact, effectiveness means the degree to which objectives are achieved and the extent to which targeted problems are solved.

And therefore, the effectiveness of emergency response plan can be defined as the degree to which the emergency plans executed right by the decision makers are achieved, or the extent to which the emergency disaster risk to reduce and rapid response rescue.

By means of the definition of the emergency response effectiveness, we can put forth a comprehensive index to measure the degree of the emergency response.

Emergency response effectiveness concept includes: 1) it is a comprehensive index; 2) given a specific emergency conditions and specific mission, which degree does the emergency response plan perform in the mission.

According to the concept of the emergency response effectiveness, we can measure the emergency response plans of the earthquake disaster in some aspects including: rescue speed, capability, safety and cost. As we know that, judge whether a well earthquake emergency plan is valid need comprehensive evaluate the above factors.

3.2. Evaluation indices systems for the emergency response preparedness

As a comprehensive effectiveness index, it needs to decompose into detailed sub indices. Using the top-down analysis methodology of system engineering, we can break up the top effectiveness indices into the different sub-indices, seeing in the Figure 2.

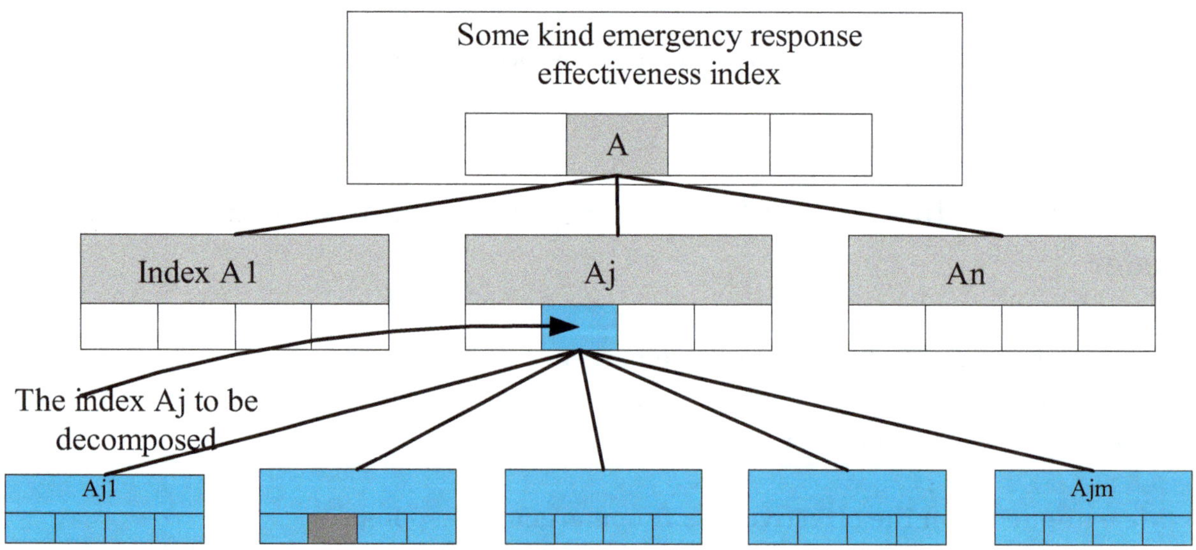

Figure 2. Effectiveness indices decomposed in top-down methodology

With the top-down methodology, we can design the response effectiveness indices system for the untraditional earthquake disaster, for example, seeing Figure 3.

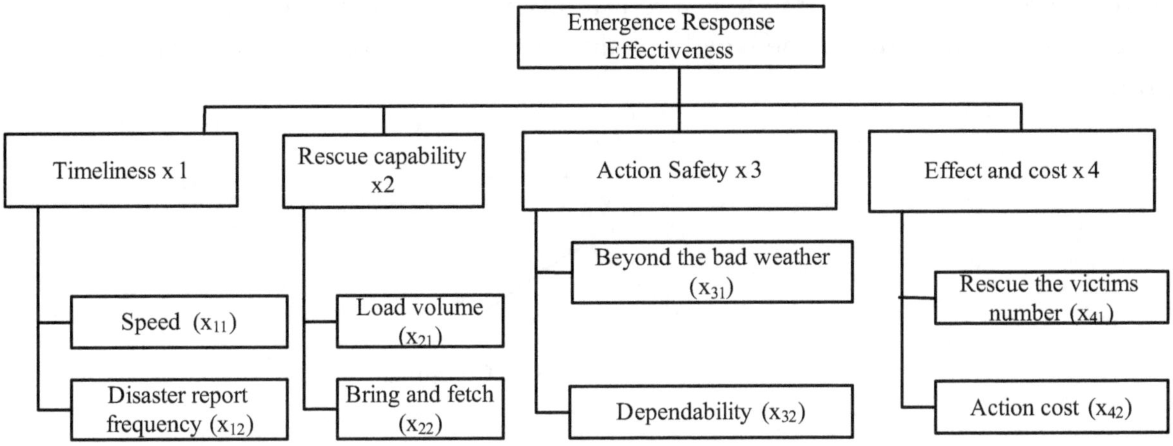

Figure 3. Response effectiveness indices of earthquake disaster

In order to describe the indices clearly and be easy to design for programming, we adopt the XML (Extensible Markup Language) file format to describe the effectiveness indices of the earthquake disaster. The emergency response effectiveness indices system can be written in XML file format as follows:

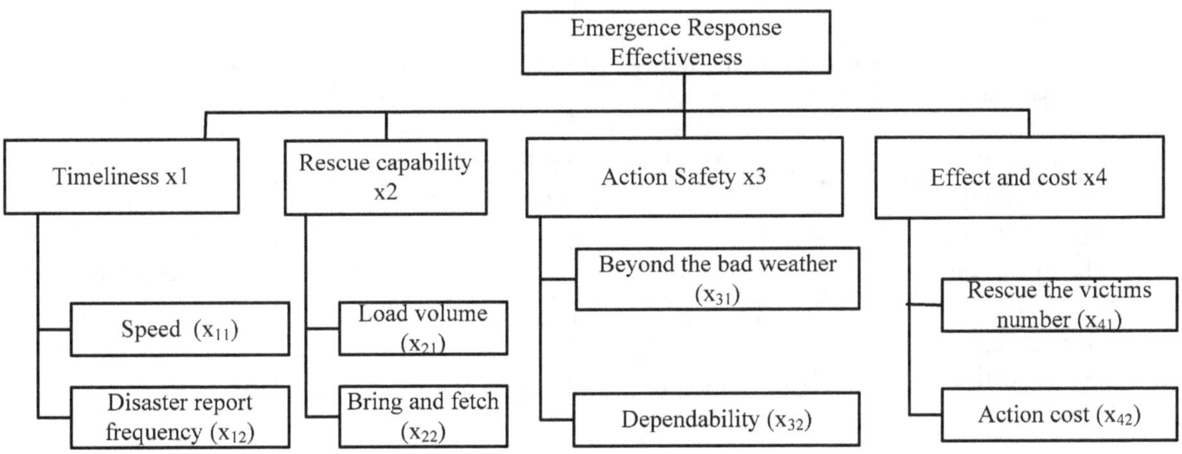

Figure 4. The emergency response effectiveness indices system

3.3. The weights acquirement for the emergency response effectiveness indices

For the evaluation indices system, we have to know how much the sub-indices make role in evaluating the emergency response preparedness. And therefore, it is essential to acquire and calculate the weights of the sub-indices, which support the whole response plans effectiveness. However, it is complicated to acquire the indices weights because the calculation full of experts' interaction with the evaluation systems.

In order to calculate the indices weights smoothly and make such an weights acquirement system reusable, we design a weights evaluation system based on MVC (Model View and

Controller) mode. The MVC mode is a reusability mode in software engineering, seeing the Figure 5. The architecture of the weight evaluation system based on MVC is designed as the Figure 6.

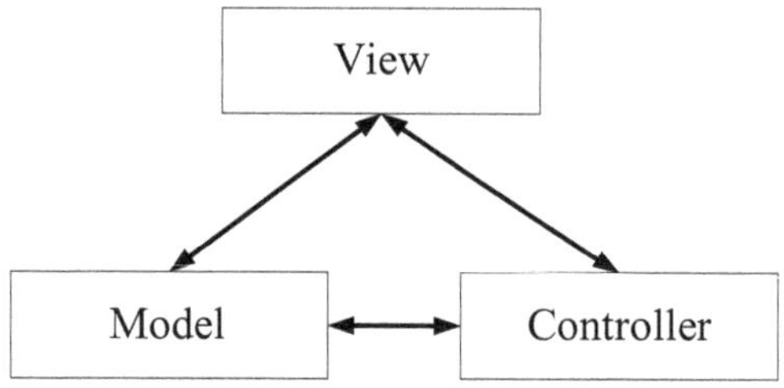

Figure 5. The Model-View-Controller mode in software engineering

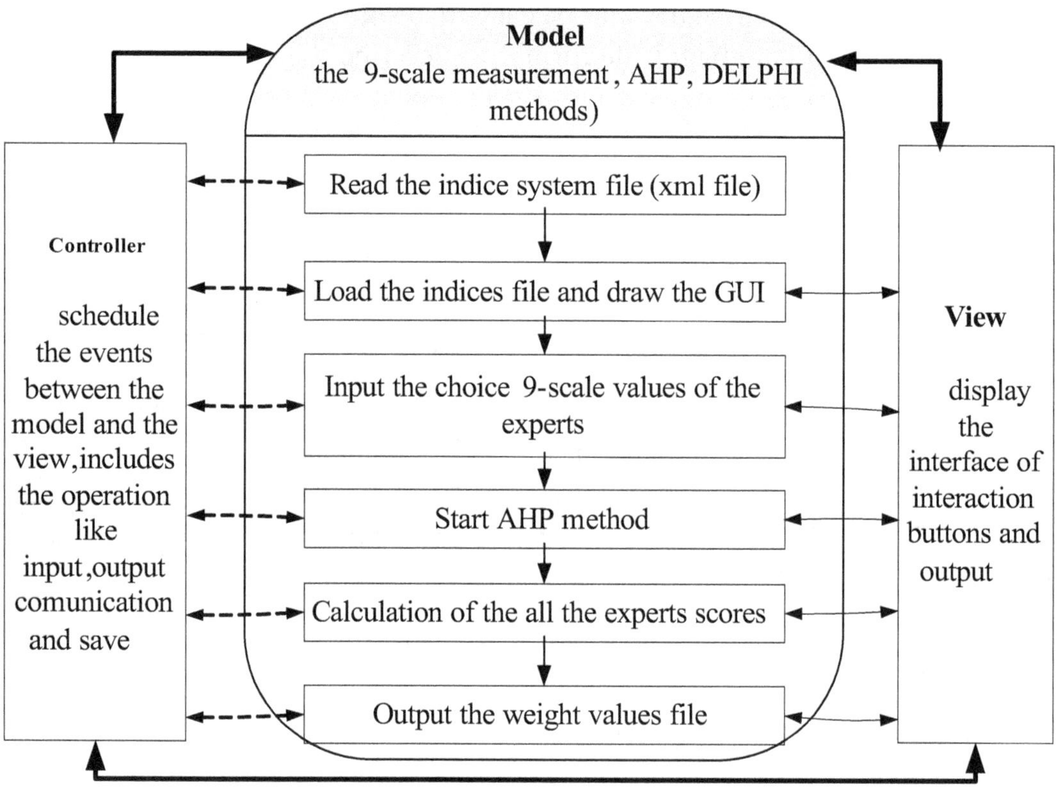

Figure 6. The architecture of the weight acquisition system based on MVC

In the Figure 6,the main parts are described as follows:

Model: it is a process algorithm of expert scoring information based on the idea of AHP (Analytic Hierarchy Process). Before the model starts, the response effectiveness indices system file (xml file) will be loaded in the weights acquisition system.

View: it is the graphic user interface (GUI) and it displays the visual information including the input windows of the experts, and the print of the weight values output.

Controller: it is the communication and control bridge between the Model and View, and it mainly includes the operations of the input, output, and calculation and saving.

As an application example, we load the xml file of the response effectiveness indices (seeing the response effectiveness indices system, Figure 3.), then the weight evaluation system with the graphic interface view is generated, (seeing the Figure 7). This system is developed in java programming language. Notice, Here the indicators are the same meaning of indices.

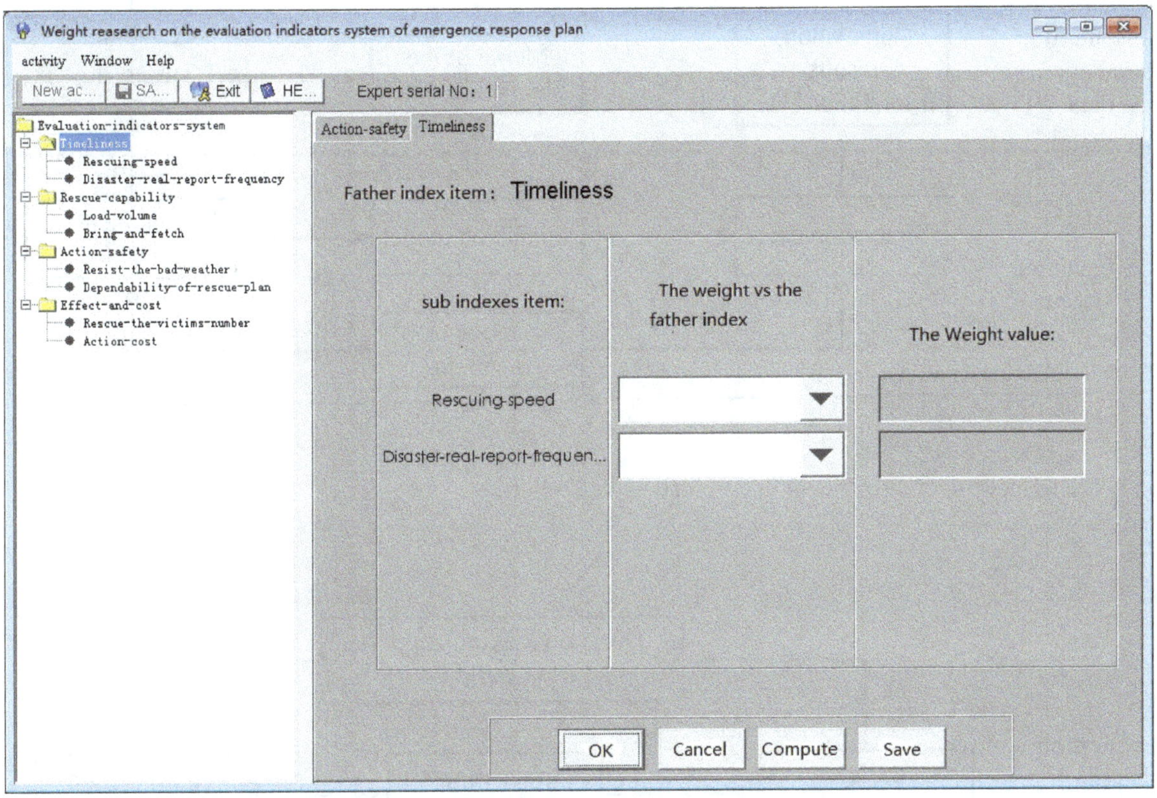

Figure 7. The weights calculation and the visual interaction system

By means of above system, we can get the indices weights W from index x_1 to x_8 : W=(0.142, 0.115, 0.091, 0.10, 0.092, 0.123, 0.159, 0.178). The weights W will be used in the integrated evaluation in later of this chapter.

3.4. The acquirement of the simulation information

The response plans of the emergency event are designed for the future emergency, so that we cannot collect the real data of emergency event, and therefore the simulation method are necessary to evaluate the future emergency plans. However, it is difficult to get the proper information from the simulation system because there is much difference between the simulation and evaluation. As we know that, the simulation system cannot get the proper information without the evaluation indices requirements. So we try to build a bridge between the two subsystems. This bridge is the so-called framework of menu of simulation data collecting, seeing the Figure 8.

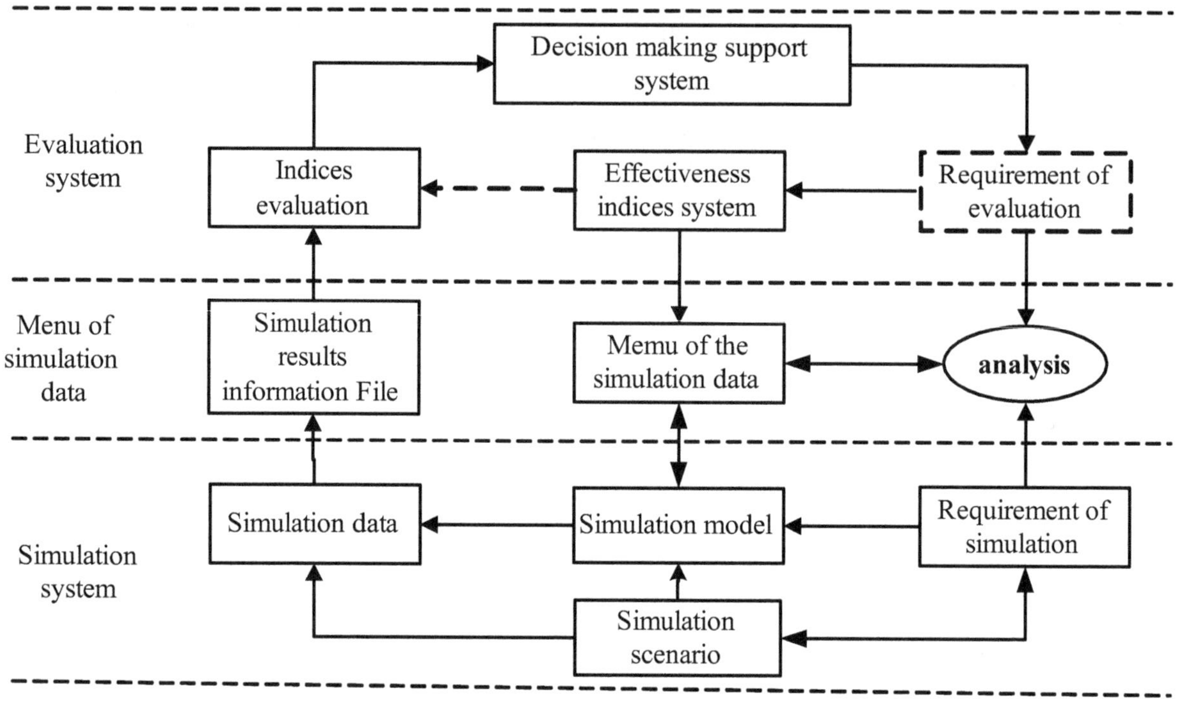

Figure 8. The framework of the menu of the simulation data collecting

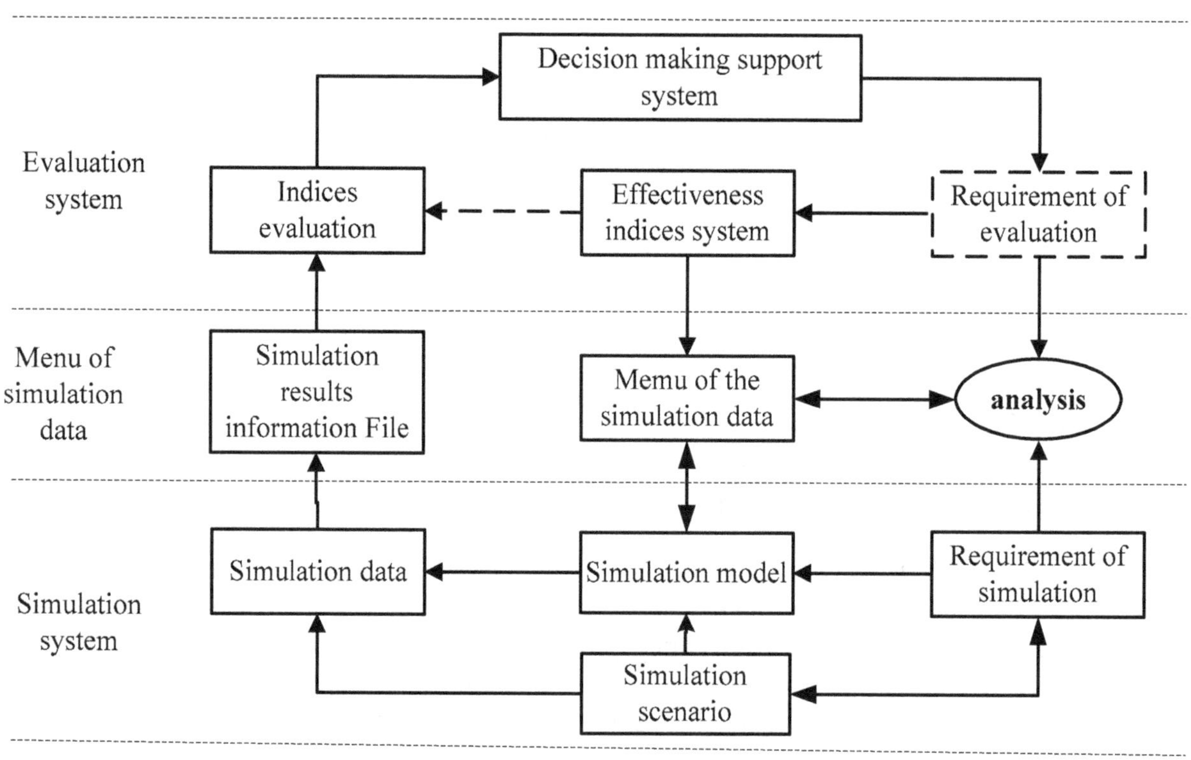

Figure 9. The menu of the simulation data

Because the simulation data is isomerous, it is difficult to save and transfer in form of the relational database format. In accordance with the format of emergency response effectiveness indices, the proper file format for the menu of the simulation data is the XML file. For the earthquake emergency response scenario, we design the menu of the simulation data to collect the simulation information as follows, seeing in Figure 9.

4. Simulation mechanism of the emergency response process

As we know that, it is difficult to evaluate the emergency response effectiveness of the earthquake disaster in real environments. It is necessary to use simulation method to support the decision making. And therefore, a reasonable simulation method should be based on a scientific mechanism of emergency simulation.

4.1. The mechanism on the emergency response process of an earthquake disaster

In order to deal with the emergency response to the earthquake disasters, the government especial organization such as emergency departments and their executers should keep reconnaissance on the disaster zone, and make actions according to the detail disaster conditions.

How to describe such rules of the emergency process is a big problem. Considering the OODA (Observe, Orient, Decide, Act) loop which is John. Boyd, the famous U.S. captain, first put forth. It tells us the combat rule which runs as a loop, seeing as Figure 10.

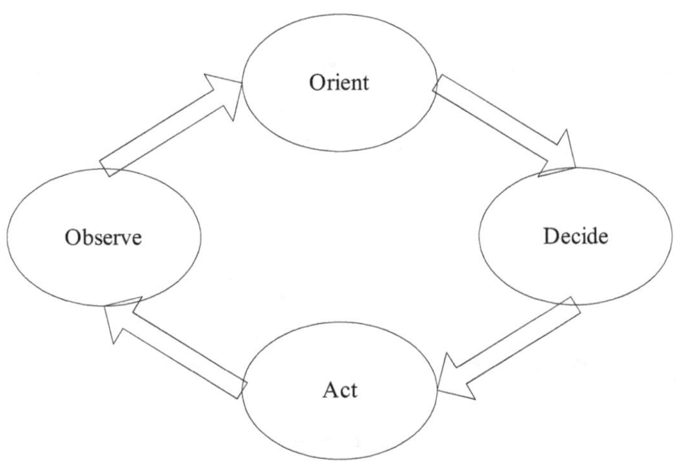

Figure 10. OODA loop diagram

In fact, the rapid response to the earthquake disaster is similar to a combat. And therefore, we should build the emergency response framework based on OODA loop as follows, seeing the Figure 11.

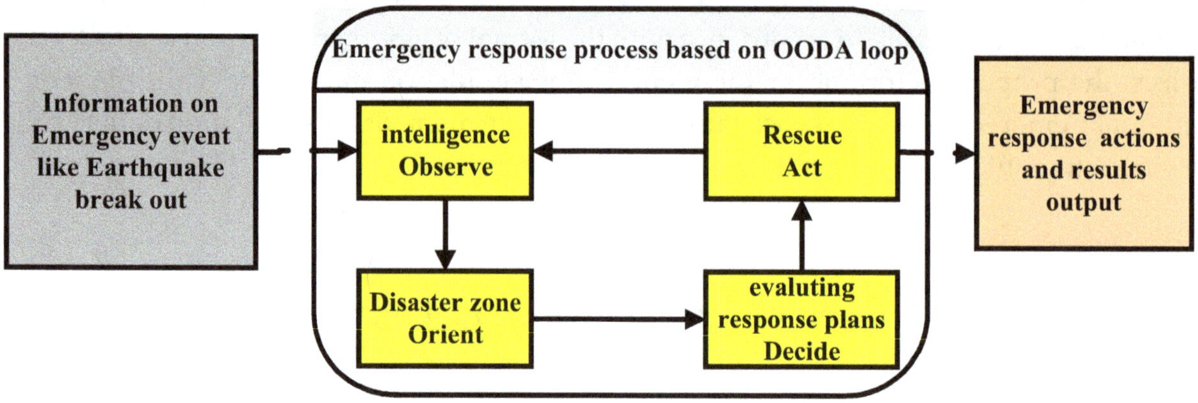

Figure 11. The response framework is based on OODA loop for the earthquake disaster

4.2. Emergency response modeling theory based on OODA

With the emergency response framework for the earthquake disaster, we need build a simulation model in scientific language. DEVS (Discrete Event System Specification) is a formalism language for modeling and analysis of discrete event systems (DESs). The emergency response system just is such a system, so that it is reasonable to using DEVS to describe the simulation process.

4.2.1. The OODA loop and the DEVS theory

The DEVS formalism was invented by Bernard P. Zeigler, who is emeritus professor at the University of Arizona. It is a modular and hierarchical formalism for modeling and analyzing general systems. There are two key models: atomic DEVS and coupled DEVS. *Atomic DEVS* captures the system behavior, while *Coupled DEVS* describes the structure of system.

The atomic DEVS is fit for describing the emergency response system behavior. Combining with the emergency response process framework based on OODA, we can build up the control structure in the DEVS factors, seeing in Figure 12.

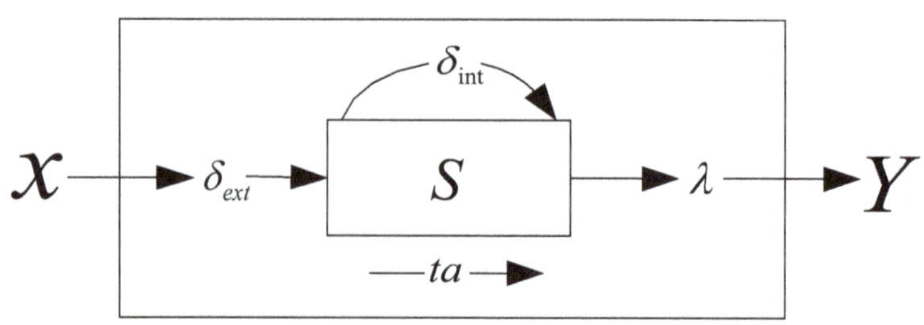

Figure 12. DEVS atomic model diagram

An atomic DEVS model is defined as a 7-tuple

AtomicDEVS= $<S,\ ta,\ \delta_{int},\ X,\ \delta_{ext},\ Y,\ \lambda>$

Where:

S is the set of sequential states (or also called the set of partial states);

X is the set of input events;

Y is the set of output events;

$Q=\{(s, t_e) \mid s \in S, 0 \le t_e \le ta(s)\}$ is the set of total states, and t_e is the elapsed time since the last event;

$\delta_{\text{int}}: S \to S$ is the internal transition function which defines how a state of the system changes internally (when the elapsed time reaches the lifetime of the state);

$\delta_{ext}: Q \times X \to S$ is the external transition function which defines how an input event changes a state of the system.

$\lambda: S \to Y$ is the output function. This function defines how a state of the system generates an output event (when the elapsed time reaches the lifetime of the state);

The output function $\lambda(s)$ is decided by the state s before transition.

$ta: S \to R^{+}_{0,\infty}$ is the time advance function which is used to determine the lifespan of a state, $R^{+}_{0,\infty}$ means Non negative real numbers; when no input events arrive, the state s will maintain the lifetime. When $ta(s)=0$, we call the state s is executing state; while $ta(s)=+\infty$, the state s is called dead state, and the system always waits for the input event.

4.2.2. Describe the emergency response process in OODA-DEVS

The kernel of a simulation system is to the simulation engine. The simulation engine of the emergency response system is required to design based on the DEVS simulation theory. Considering of the emergency response process is an OODA loop, the simulation engine of the emergency response system for disaster is designed as follows, seeing the Figure 13.

In the Figure 13, it tells us the workflow of the system simulation engine. When simulation start at $T_{0,}$ each variable in the simulation system is initiated. Then scan the current simulation time t, and judge whether the t is more than the total simulation time T. If t>T, then the emergency simulation process is terminated, else it requires to judge the emergency events. Before dealing with the messages of the emergency system, it is needed to judge the kinds of the messages. 1) If the received message is the input event message-X-message, what's more, the simulation time t of the input event must meet the requirement of the next event time, we can define such an event as the ex-message; if the simulation time of the input event is wrong, show that the synchronization time is invalid, and the event must be cancelled and to be scanned again.

After confirm the ex-message, it is required to update the simulation time of last event t_L and the time of the next event t_N, and then output the simulation result information by means of the output function$\lambda(S)$. While the response units like rescuer teams send the simulation information to the top level simulation unit (decision making departments), the response units

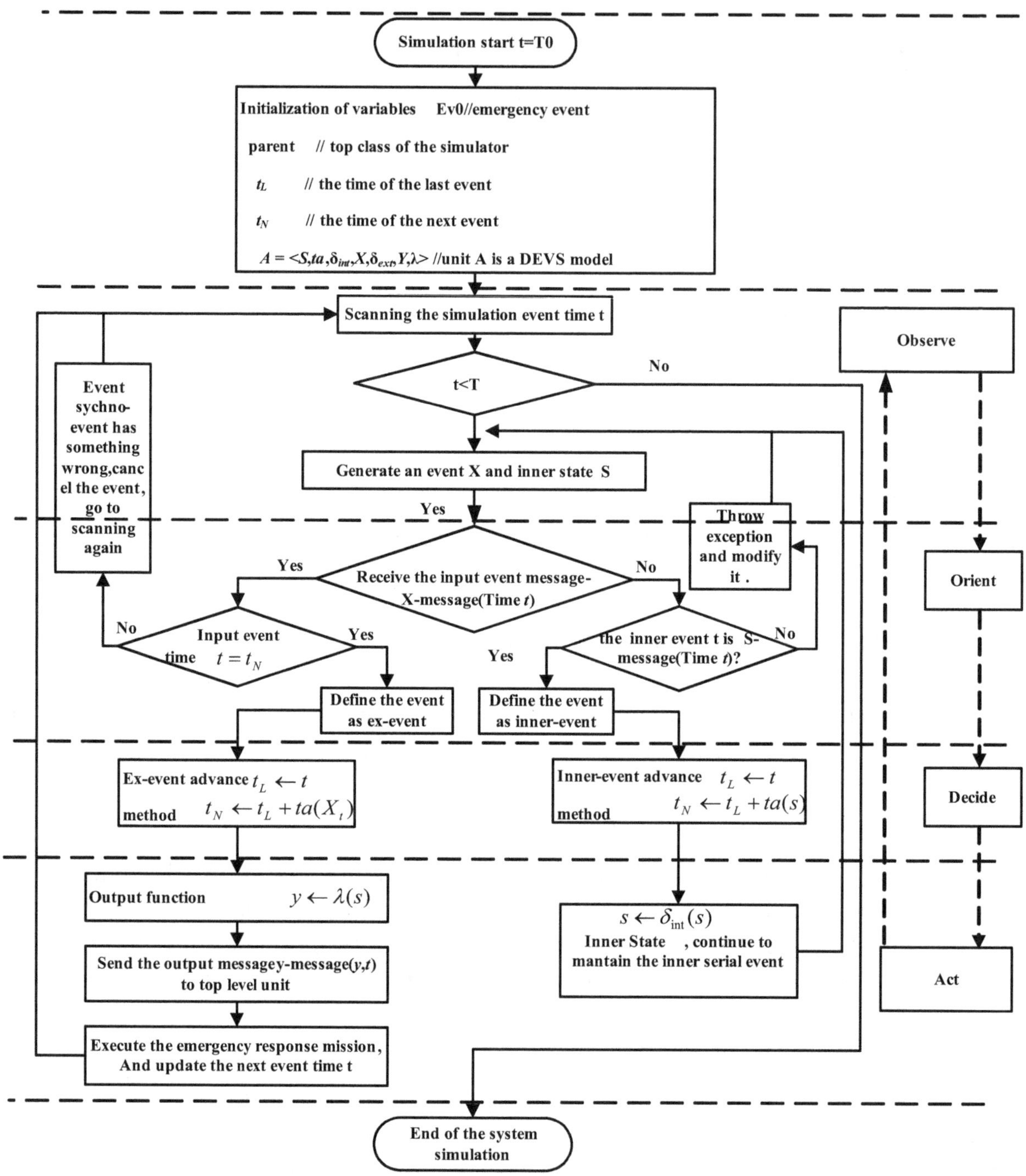

Figure 13. OODA-DEVS simulation engine

themselves will continue to finish their missions. Such units is required to update the simulation time t again, and return to scan the simulation event time t. so far, an external input event is finished dealing with.

2) If the message received is an inner message, then we define it as an inner serial event. Other wise, confirm it as an exceptional message, and it needs to modify in the time management (a simulation service rule). For the inner serial event, the inner state transits, and update the time

of the last event, and the next event time t. continue to these steps till the ex-message is coming, and transfer to the ex state loop. From above flow chart analysis, it is easy to find out, the simulation engine process just like the OODA four parts loop.

4.3. Simulation system and simulation data list acquisition

4.3.1. The simulation architecture of the emergency response management

As we all know that, the earthquake emergency response system is very complex. The simulation system includes many parts: the emergency scenario, the behaviors of response units, the interaction communications between the rescuer and the refugees. All the parts are based on the OODA loop in the response process.

Considering the emergency response plans need the many interaction operations, so that we build up a simulation system architecture based on the HLA/RTI (High Level Architecture/ Run-time Infrastructure) technology. It is suitable for the modeling developers to use HLA and RTI to describe emergency system with many uncertainty events.

The simulation architecture includes many software nodes: response plan and rescue units, the emergency information like weather and location, situation description, the environment of disaster area, emergency events generator, and rescue command system, and so on.

The simulation architecture of the emergency response for earthquake disaster is shown in Figure 14.

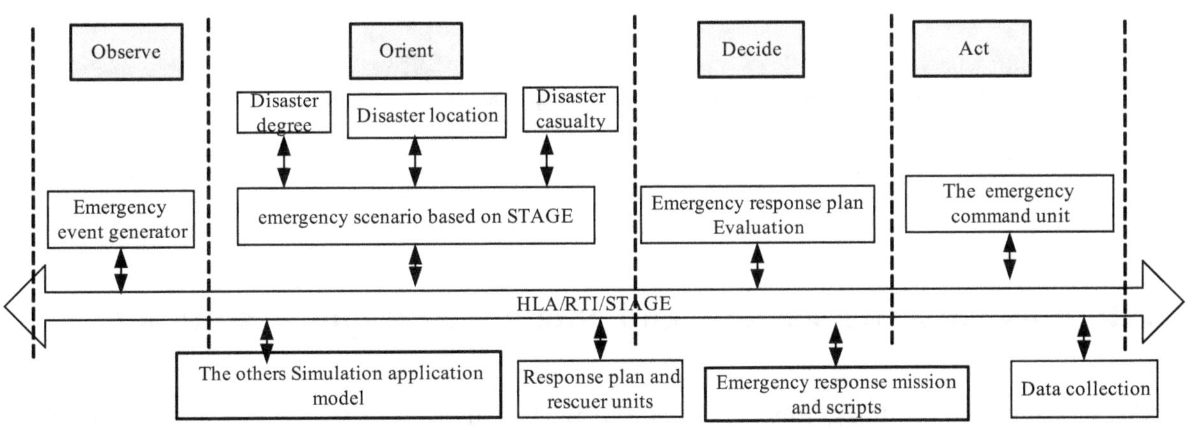

Figure 14. Simulation architecture of the emergency response

4.3.2. Response units behaviors simulation scripts

After we build a simulation system, we need analyze the simulation behaviors. In the simulation toolkit of STAGE (Scenario Toolkit and Generation Environment), we can create the behavior simulation scripts based on the above simulation theory and the simulation engine.

Emergency response unit is the force to execute the rescue mission like troop, emergency medical services, police, and firefighters and so on. For each of the emergency response unit, there is a script program to describe the unit behavior. The script codes are written in STAGE as follows.

```
(1) The script codes for detection of the emergency response units
    int detect_platforms[500];                //the numbers of detect platforms
    track.cycle_on(detected_platforms);       //start scanning the detect platforms
      while((track.next()>0)and(track.indentify=danger)and(track.type=land)) do
      detect_platforms[index]=1;   //detected the emergency response units, its value is 1.
    endwhile
(2) The orient script code of the emergency response unit
    Boolean detected[500]; int i=l;boolen identify; //initiating the variables
    enum {best, better, good, worse, worst} Degree;// the enum type about disaster degree
    while(i<500)do
     if (identify)
        detected[i]=1;
      else detected[i]=0;
    while(detected[i]!=0)do
      object.orient(i,Degree);
      data_link.assign_int(i);
      i=i+1;
    endwhile;

(3) The decision and action of the emergency response unit:
    int act_index; //the index which need to rescue
    track.cycle_on(act_index);                //scan the act_index
    while(track.next()>0)do           //there are objects required to rescue
    if(boolen rescue==1 and the orient[i]==worse)then
     entity.request_ground_track=1; // require ground tracks to rescue.
     entity.request_speed=entity.actual_speed+entity.max_acceleration*10; //require the speed of the
rescue truck
    endif;
      if(rescue==0)then
         entity.request--speed=0;  //the emergency response unit is stopping
      endif;
    endwhile;
```

5. Earthquake disaster scenario and emergency response simulation

In order to analyze the emergency response plans of earthquake disasters, a typical earthquake disaster is designed as an application scenario. The next emergency response plans are evaluated based on such a scenario.

5.1. Earthquake disaster scenario

Suppose a terrible catastrophe like earthquake breaks out in some uncertainty place, seeing the Figure 15, we should carry out the emergency response plans to quickly rescue the victims. Usually the routes to the disaster zone are far away from city and the disaster conditions are very complicated and uncertainty, so that we are required a decision-making support system to get the best emergency plan from many available plans.

In this earthquake disaster scenario, the advantages and disadvantages of the emergency response plans can be shown in Table 1.

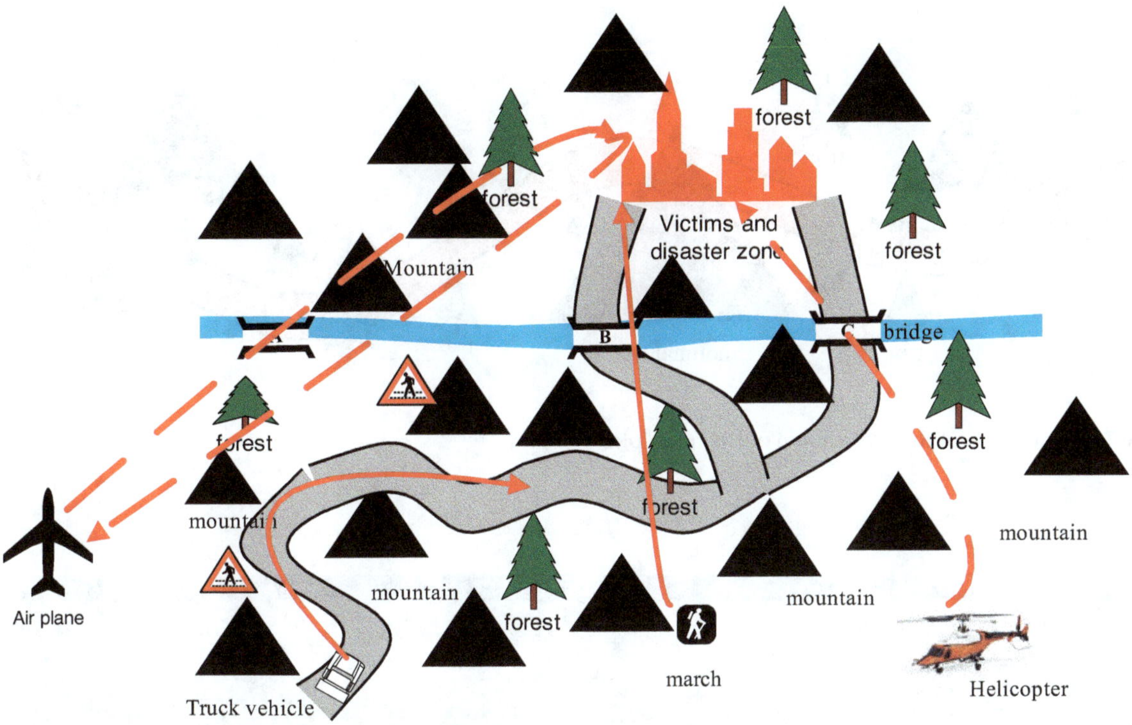

Figure 15. The Earthquake disaster scenario and some emergency response plans

Id	response plans	Advantage	Disadvantage
1	Truck Vehicle	Less relied on weather, fast and land easily.	More relied on highway traffic, less safety.
2	Air- plane	Less relied on highway, high speed and large loading capacity	More relied on weather, hard to land on disaster area and high cost.
3	Helicopter	Less relied on highway, high speed, good flexibility, take off and land easily.	More relied on weather, less safety and high cost.
4	March	Less relied on highway, good flexibility, flexible route	Carry less and move slowly.
m	Plan M	......	......

Table 1. The advantages and disadvantages of some emergency response plans

5.2. The acquisition of the simulation data

For example, we select the emergency response plan by truck vehicle to simulate. In this plan simulation, first to arrange vehicles direction to the disaster zone, and then to design the routes with the random disasters conditions based on the discrete event simulation method (see Figure.16). During the simulation time, it is possible that the rescue trucks are hit by roll rock or mudslide from the secondary disasters, and it will spend much time to deal with the breakdown. We can simulate the trucks rescue action in different routes in the STAGE (see Figure.17).

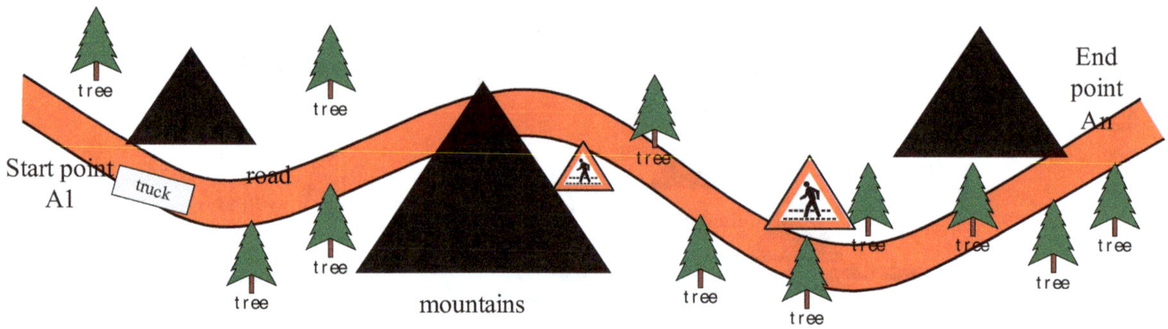

Figure 16. Simulation of the breakdown of the truck plan

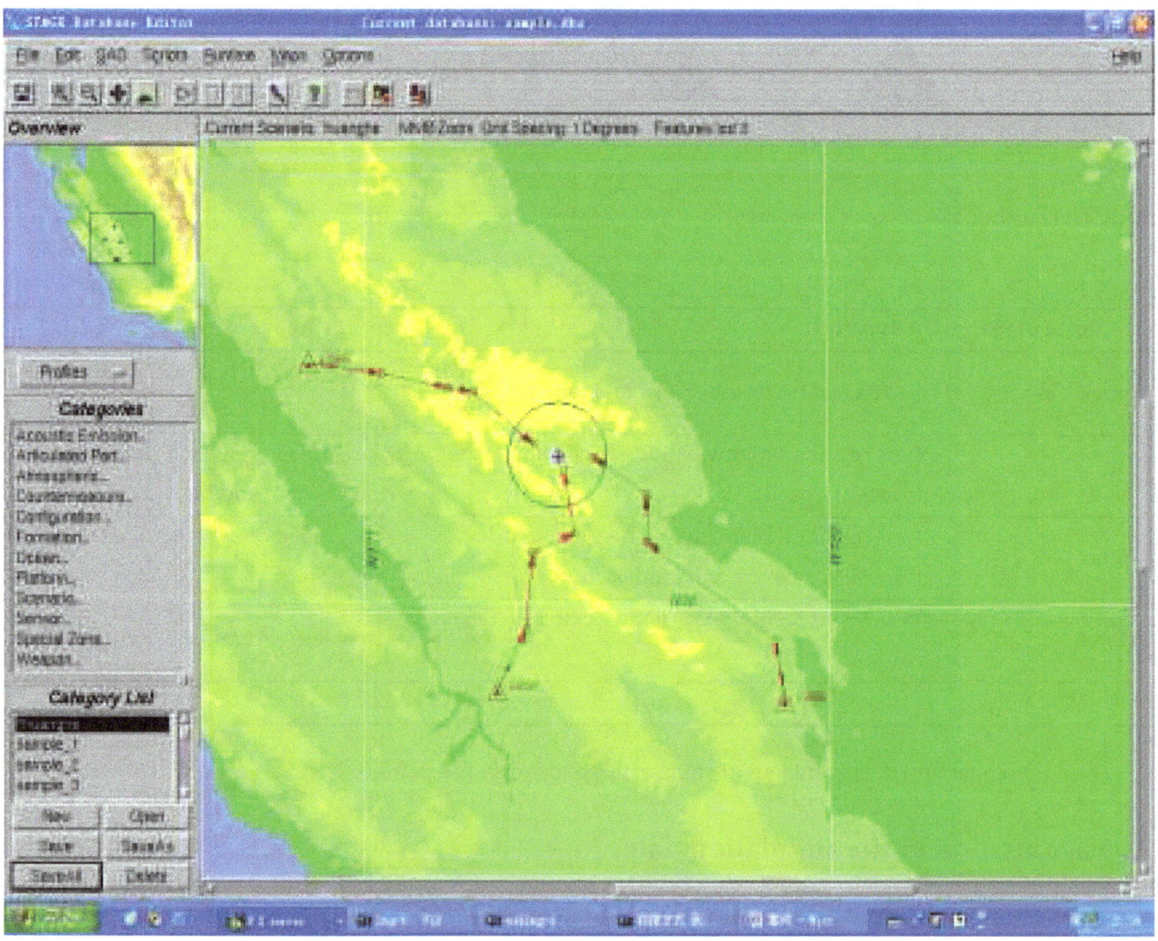

Figure 17. Simulation on the process of truck plan

Through the simulation system, we can collect the needed simulation information and save it as Xml file. The simulation information file (Truck.xml) is written in the format of the menu of simulation data as follows, seeing Figure 18:


```
<?xml version="1.0" encoding="UTF-8" ?>
< Evaluation-indices-system-of-emergency-response-plan >
    <Timeliness >
        < Rescuing-speed start-time=" 0" arrival time="4" distance=" 300.0"/>
        < Disaster–realtime-report frequency report –times-about-the-disaster="15"/>
    </ Timeliness >
    < Rescue-capability >
        < Load-volume available-weight-to-carry -and-fetch="12"/>
        < Bring-and -fetch equipment-capability-to-carry-and-fetch="0.6" />
    </ Rescue-capability >
        < Action-safety >
            < Beyond-the-bad-weather Level-of-last-bad-weather="6"/>
            < Dependability-of-rescue plan Dependability-probability="0.7"/>
    </ Action-safety >
    < Effect-and-cost >
            < Rescue-the-victims number Number-being-rescued="10"/>
    < Action-cost Cost-of–the-rescue-plan="10"/>
    </ Effect-and-cost >
</ Evaluation-indices-system-of-emergency-response-plan >
```


Figure 18. Simulation information in Xml file

The Truck.xml file is the simulation results information about adopting the truck plan in the emergency response. The simulation and data collection of the other response plans are similar to the plan of the truck. Here it is not necessary to repeat the other emergency plans.

6. The integrated evaluation and decision making for the emergency plans

As a decision-making system for the emergency response plans, it is necessary to integrate the above simulation data and the weights information of the evaluation indices system from the experts.

As for the evaluation method, because the evaluation indices are the isomer information which has different attributes. Therefore, in this paper, a good general evaluation model is adopted to build the decision-making method.

The detail evaluation method is researched as follows.

6.1. Build the evaluation indices system

The effectiveness evaluation indices system is built in Figure 19. A_i is the index i, which describes the system attributes of the alternatives. Among the indices, there are three different types of indices: one type is the effective index, which value is the more, the better; while the second type is the cost index, which value is the more, the worse; and the third type is the proper index, which characteristic is between the two indices.

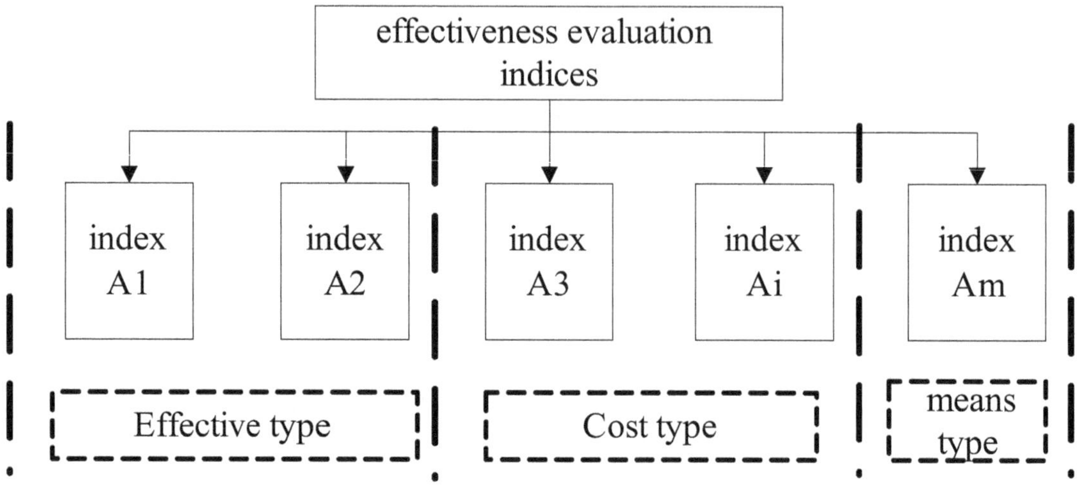

Figure 19. The general evaluation indices system on the emergency effectiveness

6.2. Effectiveness evaluation model based on TOPSIS

TOPSIS is for short of the Technique for Order Preference by Similarity to Ideal Solution. Its main idea is to evaluate the alternatives by means of the measurement scale named Euclidean distance. It is fit for evaluating many alternatives, which have different kinds of attributes.

The evaluation method based on the TOPSIS is very suitable for evaluating the effect of the alternatives, which have many heterogeneous indices. In fact, this method is built on a decision-making matrix, which different rows in the same column index have the same attribute. This is the reason that the heterogeneous data from different plans can also be compared in this evaluation model.

First all, it is necessary to collect the related evaluation data. Combining with the evaluation indices system, get the data or attribute values from the different evaluation indices. Using such data or attributes values, we can build the evaluation matrix C, seeing the Table 2. In this table, the row denotes the alternatives, and the column name stands for the attributes values of the indices. Here the attributes data is from simulation information.

alternative	attribute 1	attribute 2	…	attribute n
alternative1	C_{11}	C_{21}	…	C_{1n}
…	…	…	…	…
alternative m	C_{n1}	C_{22}	…	C_{mn}

Table 2. Alternative-attribute information

C_{ij} means the j-th attribute of the i-th alternative or emergency plan. Where i=1,2,…,m; j=1,2, …,n.

The evaluation method based on the TOPSIS is describes as follows.

Firstly, convert the attributes in table 2 into the decision matrix C.

$$C = \begin{pmatrix} c_{11} & c_{12} & \cdots & c_{1n} \\ c_{21} & c_{22} & \cdots & c_{2n} \\ \cdots & \cdots & \cdots & \cdots \\ c_{m1} & c_{m2} & \cdots & c_{mn} \end{pmatrix}$$
$$r_{ij} = w_j c_{ij} \Big/ \sqrt{\sum_{i=1}^{n} c_{ij}^2} \tag{1}$$

For the data from the matrix C, the attributes from the same column can be compared with different rows. In order to be convenient to evaluate the different alternatives, we need to standardize the matrix C into matrix R with the unit r_{ij} in formula (1). For w_j, it is the j-th weight of the indices. The matrix R becomes a normalization matrix which also adds the related weights from the evaluation indices, and R will be the foundation of the next evaluation steps.

Secondly, analyze the ideal point and negative-ideal point from the alternatives.

To get the ideal alternative and negative-ideal alternative, it depends on the type of the indices. Different type indices have different modes of process. The ideal alternative vector point is x* in the formula (2). At the same time, the negative-ideal alternative vector point is x-in the formula (3):

$$\{(\max_{i} r_{ij} | j \in J), (\min_{i} r_{ij} \mid j \in J') \mid i \in M\} = [r_1^*, r_2^*, \ldots, r_n^*] \tag{2}$$

$$\{(\min_{i} r_{ij} | j \in J), (\max_{i} r_{ij} \mid j \in J') \mid i \in M\} = [r_1^-, r_2^-, \ldots, r_n^-] \tag{3}$$

In the formula (2) (3), M is the set of the alternatives, J and J' is the effective type and cost type respectively.

Thirdly, calculate Euclidean distance of each alternative to the ideal alternative or the negative-ideal alternative.

The Euclidean distance of the i-th alternative to the ideal vector point.

$$S_i^* = \sqrt{\sum_{j=1}^{n} (r_{ij} - r_j^*)^2} \quad (i \in M) \tag{4}$$

The Euclidean distance of the i-th alternative to the negative-ideal vector point.

$$S_i^- = \sqrt{\sum_{j=1}^{n} (r_{ij} - r_j^{-*})^2} \quad (i \in M) \tag{5}$$

Fourthly, calculate the degree of each alternative approaching the ideal alternative C_i in formula (6).

$$C_i = S_i^- / (S_i^- + S_i^*) \tag{6}$$

According to the formula (6), if the degree C_i is larger, means that i-th alternative is better.

6.3. Response effectiveness indices system and evaluation information acquisition

In this application example, the effectiveness of several emergency response alternatives for the earthquake is evaluated based on the above evaluation model. The evaluation process is carried out as follows:

First, set up the evaluation indices system for the earthquake emergency response effectiveness, seeing the Figure 20.

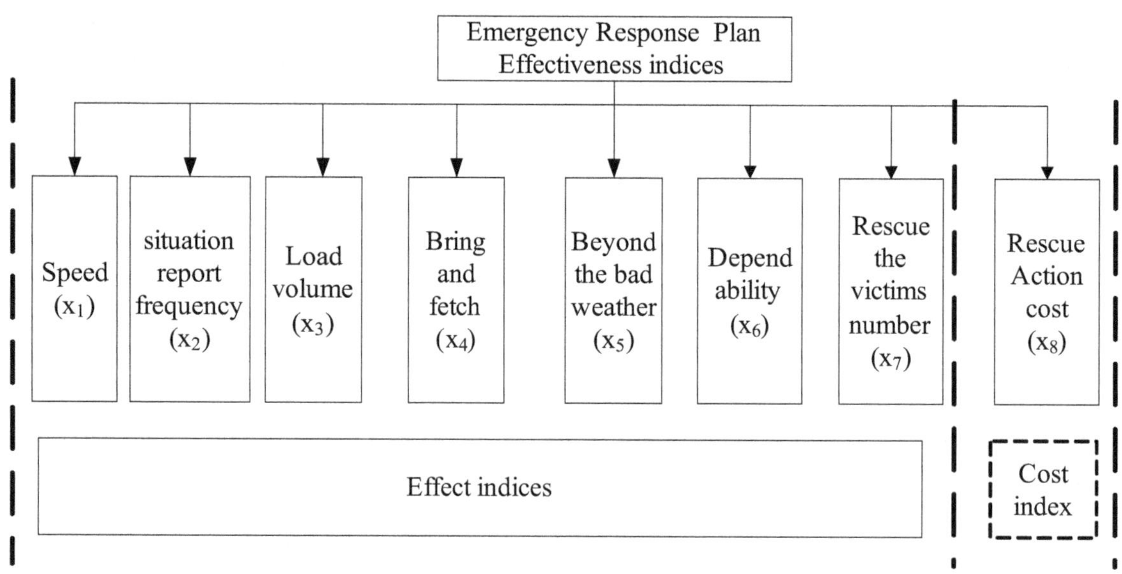

Figure 20. The effectiveness evaluation indices system

Second, there are two kinds of important information to be ready: indices weights and the indices simulation data.

According to the weights acquisition system, get the weights values from x_1 to x_8 as follows:

W=0.142, 0.115, 0.091, 0.10, 0.092, 0.123, 0.159, 0.178.

The simulation values of the indices from x_1 to x_8 are filled in the table 3, where, x_8 is the cost type index, means the less the better. The indices except x_8 are the effect indices.

Emergency response plan A	Rescuing speed x1	Disaster real time report frequency x2	Load volume x3	Bring and fetch capability x4	Beyond the bad weather x5	Dependability of rescue plan x6	Rescue the victims number x7	Action cost x8
Truck	75	15	12	0.6	6	0.7	10	10
Air plane	500	10	40	0.4	3	0.5	0.1*	40
Helicopter	300	25	20	0.9	8	0.7	20	20
march	6	20	0.5	0.6	9	0.8	1	2

Notice * : For the airplane can not carry anything back from disaster zone, here use 0.1 stands for carrying few.

Table 3. The indices simulation data of emergency response plan

6.4. Process of the evaluation

With the method, we can get the original matrix C from the information in Table 3.

$$C=\begin{bmatrix} 75.0 & 15.0 & 12.0 & 0.6 & 6.0 & 0.7 & 10.0 & 10.0 \\ 500.0 & 10.0 & 40.0 & 0.4 & 3.0 & 0.5 & 0.1 & 40.0 \\ 300.0 & 25.0 & 20.0 & 0.9 & 8.0 & 0.7 & 20.0 & 20.0 \\ 6.0 & 20.0 & 0.5 & 0.6 & 9.0 & 0.8 & 1.0 & 2.0 \end{bmatrix}$$

Due to each of the index has different weight, it is necessary to think about the indices weights for the original matrix. In this paper, the weights values for the indices from x_1 to x_8 is W=0.142, 0.115, 0.091, 0.10, 0.092, 0.123, 0.159, 0.178. Using the formula (1), we can get the standard matrix P.

$$P=\begin{bmatrix} 0.018 & 0.047 & 0.0236 & 0.0462 & 0.04 & 0.063 & 0.071 & 0.0388 \\ 0.121 & 0.0313 & 0.0786 & 0.031 & 0.02 & 0.045 & 0.0007 & 0.155 \\ 0.025 & 0.078 & 0.0393 & 0.0692 & 0.0534 & 0.063 & 0.1421 & 0.0776 \\ 0.0015 & 0.0625 & 0.001 & 0.0462 & 0.06 & 0.072 & 0.0071 & 0.0078 \end{bmatrix}$$

From the matrix P, the Ideal alternative vector point can be selected as IdealPt=(0.121, 0.078, 0.0786,0.069,0.0601, 0.072, 0.142,0.0078). Notice that X_8 is the cost index, the less the better.

At the same time, the most negative alternative point can be chosen as NegativePt=(0.0015, 0.0313, 0.0010, 0.031, 0.020, 0.045,0.00071, 0.155)

By means of the formula (4), get the Euclidean distance of each alternative to the ideal alternative vector point as follows: IdealDis=(18.93,432.2, 232.1, 67.83).

Get the Euclidean distance of each alternative to the most negative alternative vector point as follows: anti-idealDis=(77.87, 502.41, 302.18, 21.53).

Combined with the idealDis and anti-dealDis, we can get the approaching degree F which stands for any alternative approaching the ideal alternative vector point, By the meantime, far way from the most negative alternative vector point, F=(0.4932 0.3946 0.6726 0.4451).

According to the approaching degree F, rank all the response plans in descending order as follows:

Plan 3-helicopter> Plan 1-truck > Plan 4-march > Plan 2-air plane. The diagram is described in Figure 21.

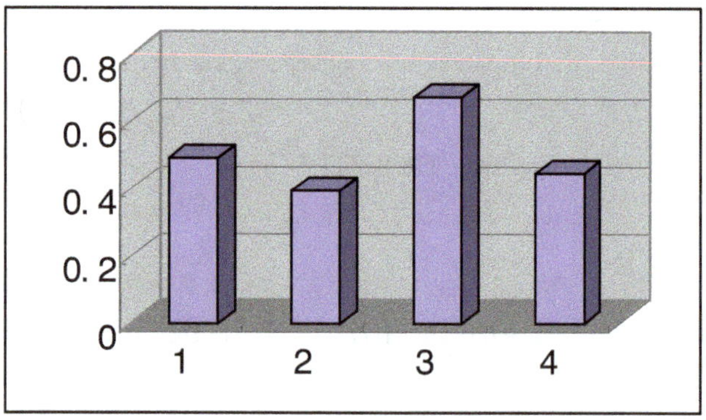

Figure 21. The effectiveness rank of the four Response plans

Based on the evaluation results, the plan 3(Helicopter rescue plan) is the first rank of all the plans, and the truck vehicle plan is the second. However, the last is the airplane response plan because it could not make its function to the maximum for example the plane can not land on the damage airport and pick up nothing back.

6.5. Future work

In this chapter, we develop a robust decision making support system, what's more, make valid application for evaluation and simulation the emergency response to the earthquake. As far as the future work concerned, there are several points to be done:

Firstly, improve the visual system of decision making system. A good visual system is welcome because we prefer the diagram to text or formulas. The future work will enforce the visual modular in the decision making system.

Secondly, improve the simulation technology for analyzing the emergency response plan, especially for the interaction simulation. There are many interactions in the simulation of the emergency response plans, and thus, it is necessary to enforce the Distributed interaction simulation technology like HLA/RTI.

Thirdly, improve the evaluation method. As we know, the evaluation method is the core of the decision-making support system. in this chapter, we adopt the experts judge method and the multi-attributes decision making method –TOPSIS. In the future work, we should adopt more evaluation methods to support the evaluation of the emergency response plans for the catastrophes.

7. Highlights

In this paper, there are some highlights as follows:

a. Put forth a set of systematic framework to simulate and evaluate the emergency response plans.

b. Define the concept of the emergency response effectiveness, and build up a set of the evaluation indices systems for emergency response plans.

c. Build the emergency response simulation model based on OODA, and describe it in DEVS theory.

d. Build up simulation system architecture based on HLA/RTI; collect simulation information by means of the simulation data list in XML format.

e. Build an evaluation method of emergency response effectiveness based on TOPSIS.

8. Conclusions

In this paper, put forth the concept of the emergency response effectiveness, and center on the methods of response effectiveness simulation and evaluation. A decision-making support system is built to evaluate the emergency response plans. The system is made up of several parts: indices weights acquirement subsystem, simulation mechanism of emergency response process, simulation data acquirement and the integrated evaluation method based on TOPSIS.

By means of the decision-making support system, several response plans in some earthquake disasters are evaluated for their response effectiveness. The evaluation results demonstrate that the simulation and evaluation system is available and reasonable.

9. Summary

It is difficult to judge the most suitable preparedness plan from many emergency response plans. And therefore, it is necessary to build a decision-making support system to evaluate the response plans and to select the most suitable preparedness plan.

In fact, we face with the challenge to test and evaluate the response plans under the **real conditions** of a strong earthquake disaster. Modeling and simulation methods are adopted to evaluate and demonstrate the preparedness plans for their advantages such as low cost, safety research, time-space easily converting, and so on.

In this chapter, pointing to the strong earthquake disasters, a decision-making support system framework is designed to manage the emergency response plans, and this framework includes the simulation and evaluation methods. In detail, a whole research route map is designed and developed to simulate and evaluate the response plans for the untraditional emergency disaster like earthquake. As a result, a serial of technical methods and theory models are included in the systematic framework. There are 5 main parts (subsystems) in this framework. They are including: (A) emergency response effectiveness concept and the evaluation indices, (B) indices weights acquisition,(C)emergency response simulation theory based on OODA-DEVS, (D) simulation system and simulation data acquirement,(E) the integrated evaluation process and decision making support system.

In this chapter, the concept of the emergency response effectiveness is defined, and it can use as a comprehensive index to measure the degree of the emergency response plan. What's more, the evaluation indices of the emergency response plans are built up in systems engineering method.

In order to describe the evaluation indices clearly and be easy to design for programming, we adopt the XML (Extensible Markup Language) file format to describe the response effectiveness indices of the earthquake disaster.

To get the weights of the evaluation indices, the weights evaluation system based on MVC (Model View and Controller) mode is built.

Referring as the similar process of combat, the emergency response process framework based on OODA loop is built, and put forth a simulation model in DEVS theory. By means of the simulation model, we design the flowchart of the system simulation engines.

For many interaction operations in the emergency response simulation, we build up a simulation system architecture based on the HLA/RTI (High Level Architecture/Run-time Infrastructure) technology. It is suitable for the developers to use HLA and RTI to describe emergency response systems full of interactions and interoperability.

A typical earthquake disaster is designed as an application scenario. The following emergency response plans are evaluated based on such a scenario.

It is necessary to integrate the above simulation data and the weights information of the evaluation indices system from the experts.

As for the evaluation method, because the evaluation indices are the isomer information which has different attributes. A general evaluation model based on TOPSIS is adopted to build the decision-making method.

Taking the emergency simulation scenario of the strong earthquake disaster as example, apply with the simulation and evaluation methods in this chapter, and get the evaluation results of emergency response plans. The evaluation results show feasible. In this example, the Helicopter rescue plan is evaluated as the best plan among the response plans, and the truck vehicle plan is the second. However, the last is the airplane response plan for it can not land on the damage airport and pick up nothing back.

Nomenclatures

AHP—Analytic Hierarchy Process

DEVS—Discrete Event System Specification

GUI—Graphic user Interface

HLA—High Level Architecture

MVC—Model View and Controller

M&S—Modeling and Simulation

OODA—Observe, Orient, Decide, Act

RTI—Run-time Infrastructure

STAGE—Scenario Toolkit and Generation Environment)

TOPSIS—Technique for Order Preference by Similarity to Ideal Solution

XML—Extensible Markup Language

Acknowledgements

This work is supported by National Natural Science Foundation (NNSF) of China under Grant No. 61374186. This work also gets some support from the 28th Research Institute of China Electronic Technology Group Corporation.

Author details

Yan-Yan Huang*

Address all correspondence to: imhyy@sina.com

Automation School, Nanjing University of Science and Technology, Nanjing, Jiangsu, China

References

[1] Kleindorfer, P.R. and Germaine, H.S.Managing Disruption Risks in Supply Chains, Production and Operations Management 2005; (14)1:53-68.

[2] Rosenthal, U., A. Boin, et al. Managing Crises: Threats, Dilemmas, and Opportunities. Springfield: Charles C Thomas, 2001.

[3] Yuan Hongyong, Huang Quanyi, Suguo feng,et al, theory and practice of key technologies of emergency platform system. Beijing, China: Tsinghua University,2012.

[4] Takahashi, T. Agent-Based Disaster Simulation Evaluation and its Probability Model Interpretation. The 4th International Conference on Information Systems for Crisis Response (ISCRAM2007),2007.

[5] Song Y, Hu, Y.H.,(2009),Group Decision-Making Method in the Field of Coal Mine Safety Management Based on AHP with Clustering,2009 IEEE International Conference on Service Operations, Logistics and Informatics, Chicago, USA, July 22–24, 2009

[6] Huang,Y.Y., Wang,J.Y., An Evaluation Method of Operational Effectiveness for the Emergency Plans, 2009 IEEE International Conference on Service Operations, Logistics and Informatics, Chicago, USA, July 22–24, 2009

[7] Hart, P., Rosenthal, U., Kouzmin, A., Crisis Decision Making, the centralization thesis revisited, in:Administration & Society, 1993;25(1):12-45.

[8] Van Santen willem, Jonker Catholijn. Crisis Decision Making Through a Shared Integrative Negotiation Mental Model, Proceedings of the 6th International ISCRAM Conference – Gothenburg, Sweden, May 2009

[9] WANG Fei-Yue. Web Social Media in Disaster Reduction and Emergency Management,Science & Technology Review;2008

[10] Fan Weicheng; Advisement And Suggestion To Scientific Problems Of Emergency Management For Public Incidents; Bulletin of National Natural Science Foundation of China;2007

3

Effect Evaluation of Radiative Heat Transfer and Horizontal Wind on Fire Whirlwind

Seigo Sakai

1. Introduction

More than three and a half of years ago, there were a large earthquake in north east area of Japan, i.e. East Japan great earthquake disaster [1], and the residents in the area have been suffering and still on the recovery process from the disaster. A lot of fires were observed in the northeast area of Japan, for example in Kesen-numa City, and so on. Fortunately, there were no reports of fire whirlwinds observed in this disaster, despite of a number of town area fires occurred. However, fire whirlwind is still one of the concerned accidents in the earthquake [2].

A large-scale wide area fire, such as a town area fire or a forest fire, sometimes induces a strong rotating flow, which is to be called fire whirlwind. Fire whirlwind consists of a tornado that includes flames, hot winds and sparks. One of the worst cases which should be avoided at the large-scale fire is the fire whirlwind, because the whirlwind itself is dangerous and play an important role to enhance spread of a fire due to widely scattering sparks.

Even if a small fire occurs, not only a flame induces an upwarding air flow and consumes oxygen from the neighboring, but also a current of air against the flame is present in orde to collect oxygen from wider area. Therefore, a big natural convection will be introduced in the fire current. In the case that the wind may blow from a certain direction into the fire current, uniform air suction collapses with the upwarding air flow. Consequently, a vortex is easily generated, because the fire current turns to a fire whirlwind due to accompanying the upwarding motion with rotating one. Fire whirlwind may be flowed away by a strong downstream wind, or may wander around to collect much oxygen.

There are some experimental works to investigate and reproduce the fire whirlwind in order to explore a flow property and express factors of fire whirlwind outbreaks [3-16]. There are also numerical works in order to analyze the property and the factors [17-24]. In spite that various factors, such as direction and velocity of horizontal wind or heat generation from the fire, are concerned to outbreak of a fire whirlwind,, it is not sufficient to show outbreak mechanism of a fire whirlwind.

Convective and radiative-convective heat transfer analyses with respect to fire whirlwind were also performed in our laboratory on former studies [25-27], just radiative exchange between solid surfaces was carried out. Therefore, in this study, radiative heat exchange is dealt in consideration of radiative gas using Radiation Element Method by Ray Emission Model (REM[2]) [28]. Radiative heat transfer effect on fire whirlwind is discussed.

Then, three dimensional analyses are performed to investigate the thermal and flow fields by using the Finite Volume Method [29] with introducing divergence of radiative heat flux for gas medium. The SIMPLE method is utilized to solve the discretized equations. Natural convection is caused from a plane source of constant temperature in the flat ground. Fire whirlwind is forcely generated stably just above the heat source with introducing air currents from four courners. For making of analysis models, a representative example of the fire whirlwind that occurred at Tokyo in the Great Kanto Earthquake (1923) is referred.

In the analysis, one dimensional radiative exchange analysis above the heat source is compared with three dimensional one to reduce the computational load and time. One-dimensional radiative exchange analysis is suffifient with respect to the accuracy. Then, the composition of participating gases is altered to discuss the effect of radiative heat exchange to the whirlwind flow field. Three dimensional analyses are performed to investigate the thermal and flow fields by using the Finite Volume Method with introducing divergence of radiative heat flux for gas medium. Fire whirlwind is forcely generated stably just above the heat source with introducing air currents from four courners. From the comparison of thermal and flow field between convective flow analysis and combined radiative-convective flow analysis, radiative heat exchange has a great influence to the thermal field and a less influence to the flow field. Increase of participating media concentration gives raise of temperature due to absorption and re-emission, and water vapor influences thermal field more than carbon dioxide.

2. Analysis and modelling

2.1. Radiative exchange

Consider the radiation element of participating medium, which is comprised of a polyhedron surrounded by polygons as shown in Figure 1. The spectral radiation intensity, I_λ, at $\vec{r}$ in the direction $\hat{s}$ can be expressed in terms of the radiation energy balance by

$$\frac{dI_\lambda(\vec{r},\hat{s})}{dS} = -(\kappa_\lambda + \sigma_{s,\lambda}) I_\lambda(\vec{r},\hat{s}) + \kappa_\lambda I_{b,\lambda}(T) + \frac{\sigma_{s,\lambda}}{4\pi}\int_{4\pi} I_\lambda(\vec{r},\hat{s}')\Phi_\lambda(\hat{s}' \rightarrow \hat{s})\,d\omega, \tag{1}$$

where κ_λ and $\sigma_{s,\lambda}$ are spectral absorption and scattering coefficients, respectively. Here, S is the path length in the direction $\hat{s}$, $I_{b,\lambda}$ is spectral black-body radiation intensity, $\Phi_\lambda(\hat{s}' \rightarrow \hat{s})$ is the phase function from the direction $\hat{s}'$ to $\hat{s}$, and ω is solid angle.

Considering the i-th participating radiation element, and assume that each radiation element has constant uniform temperature of T_i, refractive index and heat generation rate per unit volume, $q_{X,i}$. A ray passing and a part of the ray is scatted. The ray is consisted of absorbed, scattered and transmitted fractions. Additionally, the scattered radiation is assmed to be uniformly distributed over the element. [30]

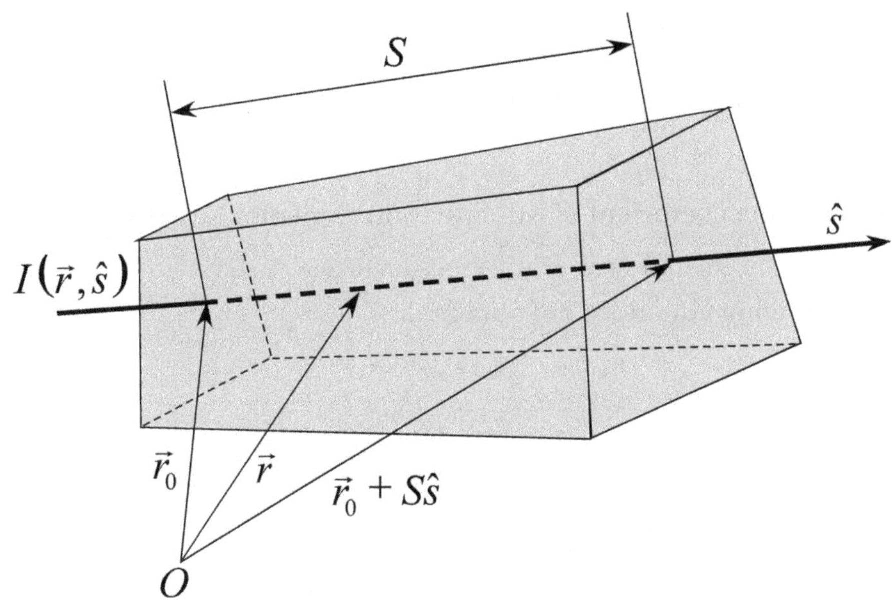

Figure 1. Radiation element

For anisotropic scattering media, an apparent extinction coefficient β_λ^* is introduced, and a corrected scattering albedo Ω_λ^* is obtained by introducing the delta function approximation [31]. Thus, an anisotropic scattering medium can be treated as an isotropic scattering one. The third term on the right hand side of equation (1) can be approximated by

$$\frac{\Omega_\lambda}{2}\int_{-1}^{1} I_\lambda(x,\mu')\Phi_\lambda(\mu')d\mu' \approx \frac{\Omega_\lambda^*}{2}\int_{-1}^{1} I_\lambda(x,\mu')d\mu' \approx \Omega_\lambda^* I_\lambda^D, \tag{2}$$

where Ω_λ is scattering albedo, μ' is directional cosine, and I_λ^D is the average scattered radiant intensity.

A radiation element i can be regarded as either a volume element or a surface boundary. Equation (1) is integrated along the path length $\bar{S}_i(\hat{s}) = V_i / A_i(\hat{s})$ and with respect to all discretized solid angles, in which V_i and $A_i(\hat{s})$ show volume of element and projection area of element onto the surface normal to $\hat{s}$, respectively [32]. The spectral radiation energy, $Q_{J,i,\lambda}$, from the radiation element i, is given by

$$Q_{J,i,\lambda} = \pi\left(\varepsilon_{i,\lambda} I_{b,i,\lambda} + \Omega^{D}_{i,\lambda} I^{D}_{i,\lambda}\right) A^{R}_{i,\lambda}, \tag{3}$$

where $\varepsilon_{i,\lambda}=1-\Omega^{D}_{i,\lambda}-\Omega^{S}_{i,\lambda}$, in which $\varepsilon_{i,\lambda}$, $\Omega^{D}_{i,\lambda}$ and $\Omega^{S}_{i,\lambda}$ are emissivity, diffuse reflectivity and specular reflectivity of i-th radiation element, respectively. $I_{b,i,\lambda}$ is spectral black-body radiation intensity of i-th radiation element, and $I^{D}_{i,\lambda}$ is average scattered radiant intensity of i-th radiation element. $A^{R}_{i,\lambda}$ is the effective radiation area [28] which is defined as follows,

$$A^{R}_{i,\lambda} \equiv \frac{1}{\pi}\int_{4\pi} A_i(\hat{s})\left[1-\exp\left(-\beta^{*}_{i,\lambda}\bar{S}_i(\hat{s})\right)\right]d\omega. \tag{4}$$

$\beta^{*}_{i,\lambda}$ is apparent extinction coefficient of i-th radiation element. By introducing the absorption view factors $F^{A}_{i,j}$ and the diffuse scattering view factors $F^{D}_{i,j}$ defined by Maruyama [31] and equation (3), the following equations are obtained:

$$Q_{T,i,\lambda} = \pi\varepsilon_{i,\lambda} I_{b,\lambda} A^{R}_{i,\lambda}, \tag{5}$$

$$Q_{J,i,\lambda} = Q_{T,i,\lambda} + \sum_{j=1}^{N} F^{D}_{j,i} Q_{J,j,\lambda}, \tag{6}$$

$$Q_{X,i,\lambda} = Q_{T,i,\lambda} - \sum_{j=1}^{N} F^{A}_{j,i} Q_{J,j,\lambda}. \tag{7}$$

In case of arbitrarily assuming the boundary conditions of the heat transfer rate of the emissive power, $Q_{T,i,\lambda}$, or the net rate of heat generation, $Q_{X,i,\lambda}$ for each radiation element, solving equations (6) and (7) using the method previously described by Maruyama and Aihara [28] gives the unknown $Q_{X,i,\lambda}$ or $Q_{T,i,\lambda}$. The relationship between $q_{X,i}$ and $Q_{X,i,\lambda}$ is obtained by

$$q_{X,i} = \frac{Q_{X,i}}{V_i} = \frac{1}{V_i}\int_0^{\infty} Q_{X,i,\lambda}\, d\lambda, \tag{8}$$

where λ is wave length. An analytical method for radiative heat transfer, i.e. the radiation element method by ray emission model, REM^2, is used in radiative heat transfer analysis, and the Statistical Narrow Band (SNB) model is combined to the REM^2 to consider the spectral dependence of the radiative properties.

2.2. Thermal and fluid flow fields

The governing equations of thermal and flow fields are the continuum equation, the Navier-Stokes equation, and the energy equation. The thermal and flow fields are assumed to be unsteady state and three dimensional. These equations are normalized and transformed to the following generalized conservation equation.

$$\frac{\partial \phi}{\partial t}+\frac{\partial}{\partial x}(u\phi)+\frac{\partial}{\partial y}(v\phi)+\frac{\partial}{\partial y}(w\phi)=\Gamma\left(\frac{\partial^2 \phi}{\partial x^2}+\frac{\partial^2 \phi}{\partial y^2}+\frac{\partial^2 \phi}{\partial z^2}\right)+S, \tag{9}$$

where t shows time, u, v and w are the normalized velocity components for x, y and z directions, respectively. Variable ϕ takes 1 for the continuum equation, u, v and w for the Navier-Stokes equation, and normalized temperature T for the energy equation. Generalized diffusion coefficient Γ takes 0 for the continuum equation, $1/\mathrm{Re}$ for the Navier-Stokes equation and $1/(\mathrm{Re}\cdot\mathrm{Pr})$ for the energy equation. Normalized source term S takes 0 for the continuum equation, the summation of the normalized pressure term and the normalized buoyancy term by Boussinesq approximation, and the normalized source term from radiative exchange mentioned in the previous section (equation(8)). Turbulent flow is treated by using high Reynolds number turbulence model.

In the thermal and fluid flow analysis, equation (9) is discretized by using the Finite Volume Method [29]. The SIMPLE method is utilized to solve the discretized equations. Physical properties of the mixture are altered depending on the change of temperature.

2.3. Analysis procedure

Figure 2 shows an analysis procedure in this study. Temporal temperature distribution is initially given to analyze nongray radiative heat transfer by REM^2. Then, the derived heat generation rate is introduced to the energy equation, and the thermal and flow field is analyzed by FVM using the SIMPLE method. The derived temperature is introduced to the REM^2 again as an initial temperature distribution, and iteration is repeated until the derived distribution is converged to the initial distribution. Steady state solution is obtained through this iteration loop.

2.4. Analytical model

Figure 3 shows analytical domain for calculation, which scale is based on the Great Kanto Earthquake (1923) in Japan. Heat source on the bottom center has 800m in width and depth, and this value is representative length L. Therefore, the analytical domain is a cubic of 2,000m in width, depth and height. Heat source is applied uniform temperature of 2,000K, and the domain is assumed to be surrounded by circumstance of 293.15K. Initial temperature of the domain is also set to 293.15K. Mixture gas is constituted by water vapor, CO_2 and Nitrogen. All the domain surfaces are assumed to be black for radiative exchange, and the surfaces except the bottom are opened. Fire whirlwind is forcely generated stably just above the heat source

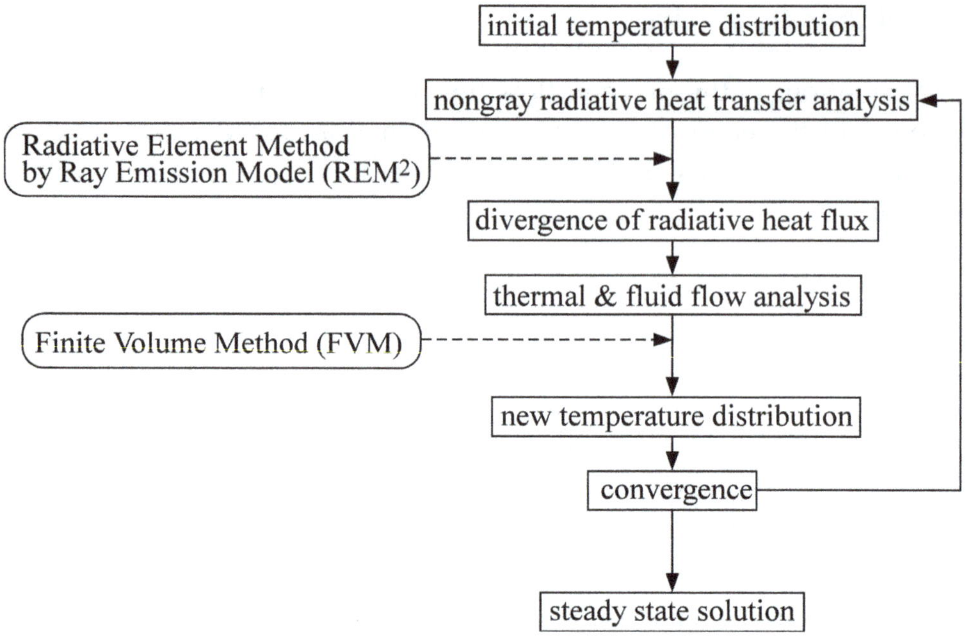

Figure 2. Flow chart of numerical simulation procedure

with introducing air currents from four corners. The currents velocities U are constant of 5m/s at the inlet surfaces of 600m in width and 200m in height. Combustion nor chemical reaction is not considered in the calculation.

Table 1 shows concentration of participating gases in mixture for radiative heat exchange. Carbon dioxide has three values; no concentration, concentration in general atomosphere and the maximum concentration in case of fire. Water vapor also has three values; no concentration, concentrarion of saturated water vapor at the initial temperature and the concentration of saturated water vapor at boiling point.

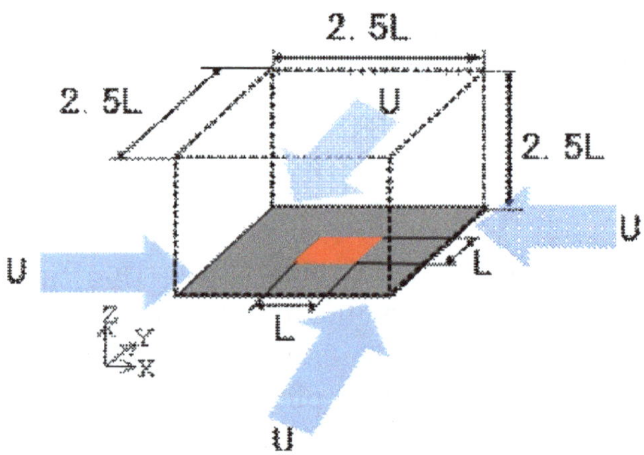

Figure 3. Analytical domain

	Concentration of CO_2 (ppm)	Concentration of H_2O (ppm)
Case A	0	0
Case B	0	3.56×10^4
Case C	3.60×10^2	3.56×10^4
Case D	8.00×10^4	3.56×10^4
Case E	8.00×10^4	1.02×10^6

Table 1. Concentration of participating gases

3. Results and discussions

In this chapter, results are shown and discussed with respect to influence of mesh spacing, effect of radiative exchange, and influence of participating media concentration on heat exchange and flow distribution.

3.1. Influence of mesh spacing on radiative exchabge and flow distribution

Before radiative exchange is coupled with convective flow calculation, influence of mesh spacing on radiative exchange and that on convective flow are individually evaluated.

3.1.1. Convective flow analysis

Figure 4 shows temperature distribution above the heat source by convective flow analysis without considering the radiative heat exchange. Mesh with uniform spacings means uniform spacings for x, y, and z directions, and mech with non-uniform spacings means uniform spacings for x, and y directions and non-uniform spacings for z direction. For all the cases, distributions are almost the same, but the coarse mesh is difficult to simulate the temperature distribution just above the heat source. Mesh with non-uniform spacings is effective to reduce computation load and to keep the accuracy.

Figure 5 shows influence of non-uniform mesh spacings on temperature distribution above the heat source by convective flow analysis without considering the radiative heat exchange. Numbers of uniform mesh spacings for x, and y directions and non-uniform spacings for z direction are increased. For all the cases, distributions are almost the same in case of higher height 400m, but result of the coarse mesh shows lower temperature distribution just above the heat source.

These two figures suggest utilization of at least 30 non-uniform mesh spacings for z direction in the convective flow analysis.

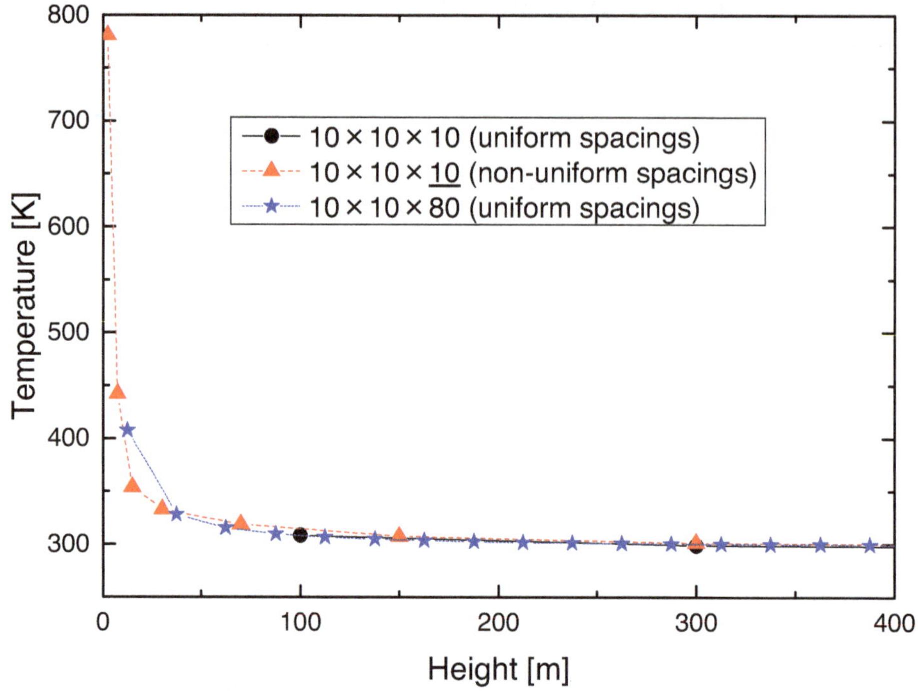

Figure 4. Influence of mesh spacings on convective flow analysis (temperature distribution above the heat source after 30 minutes from air current induction)

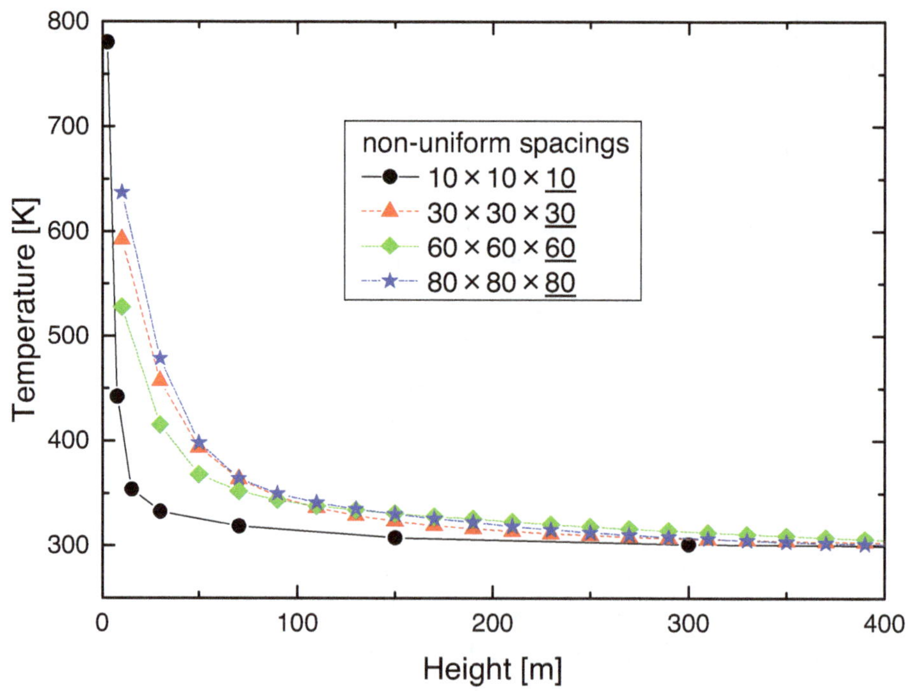

Figure 5. Influence of non-uniform mesh spacings on convective flow analysis (temperature distribution above the heat source after 30 minutes from air current induction)

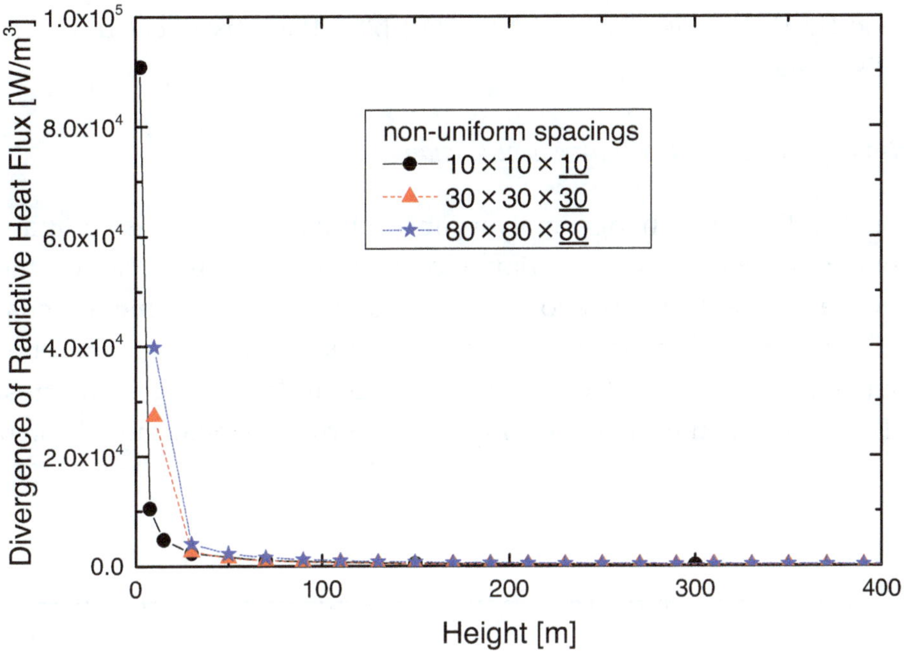

Figure 6. Divergence of radiative heat flux above the heat source after 30 minutes from air current induction

3.1.2. *Radiative exchange analysis*

Figure 6 shows divergence of radiative heat flux above the heat source. In case of coarse mesh employed, total amount of divergence of radiative heat flux is balanced with higher value just adjacent to the heat source and comparable lower value above the heat source until lower height 100m.

This figure also suggests utilization of at least 30 non-uniform mesh spacings for z direction in the radiative exchange analysis.

3.1.3. *Summary*

Individual calculation of convective flow analysis and radiative exchange suggest utilization of at least 30 non-uniform mesh spacings for z direction, and 30x30x30 mesh with non-uniform spacings is adapted for combined radiative-convective analysis.

3.2. Effect of radiative heat exchange

In our previous study, scale effect of fire whirlwind is discussed with using the numerical analysis, and relationship between a real phenomenon and the phenomenon in the reduction was examined with considering radiative heat exchange [26]. The P-1 model was utilized to simulate the radiative heat transfer from the heat source at high temperature. It was found that radiative heat exchange played an important role in the heat transfer at the higher temperature field, whereas just radiative exchange between solid surfaces was carried out.

In this study, radiative heat exchange is dealt in consideration of radiative gas using Radiation Element Method by Ray Emission Model (REM^2) [28]. Radiative heat transfer effect on fire whirlwind is discussed.

3.2.1. *Comparison of 1-d and 3-d radiative heat exchange*

One dimensional radiative exchange analysis above the heat source is compared with three dimensional one to reduce the computational load and time. Figure 7 shows the comparison of divergence of radiative heat flux above the heat source between one dimensional parallel analysis model and three dimensional analysis model. Even though the one dimensional analysis model omitted the effect of surrounding boundaries, these two results coincide comparable. Therefore, further analysis employs one dimensional model for radiative heat exchange.

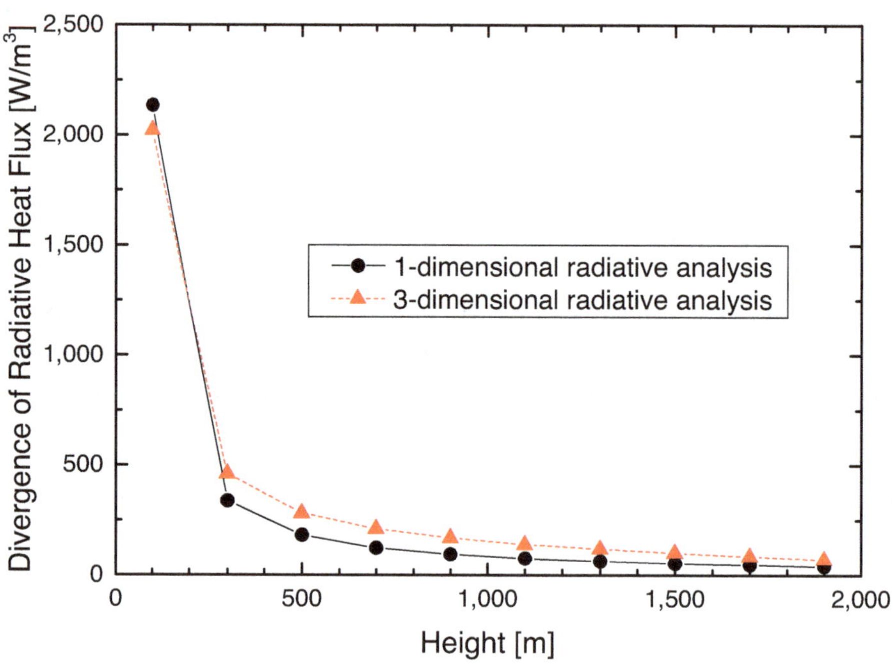

Figure 7. Comparison of 1-d and 3-d radiative analysis (Distribution of divergence of radiative heat flux above the heat source)

3.2.2. *Comparison of convective flow analysis and radiative-convective flow analysis*

Figure 8 shows heat generation rate for convective flow analysis and divergence of radiative heat flux for radiative heat exchange above the heat source. Though values of divergence of heat flux are smaller than those of heat generation rate, just convective flow analysis ignores these amounts to simulate. It is easily expected that the radiative heat exchange due to participating media plays an important role more than the radiative heat exchange between surfaces.

3.2.2.1. *Thermal field*

Figure 9 shows comparison of temperature distribution above the heat source after 30 minutes from analysis. As expected from the distribution of heat generation and divergence of radiative heat flux, tempearature distribution is different, especially until lower height 100m. Participating media have much influence to the temperature distribution, and play an important role.

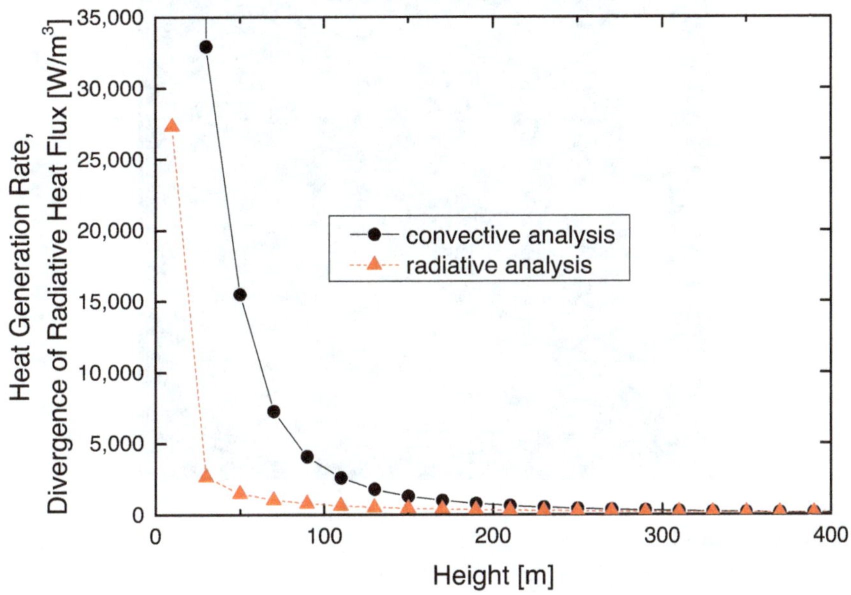

Figure 8. Comparison of heat generation rate and divergence of heat flux above the heat source

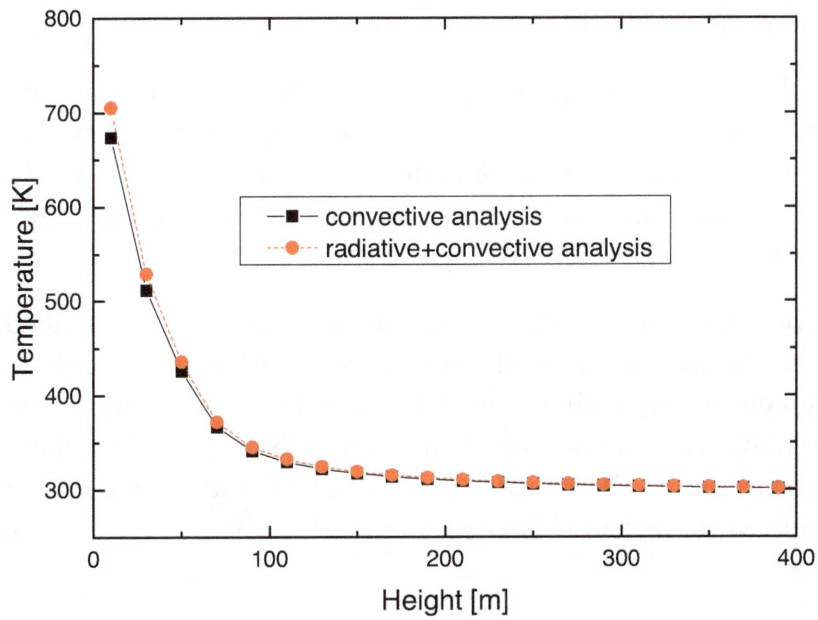

Figure 9. Comparison of temperature distribution above the heat source after 30 minutes from air current induction between convective flow analysis and combined radiative-convective flow analysis

3.2.2.2. Flow field

Figure 10 shows streamlines of the flow field. Lines are colored by velocity magnitude. Whirlwind is stably generated above the heat source, and is shrinked sharply with the height from the heat source.

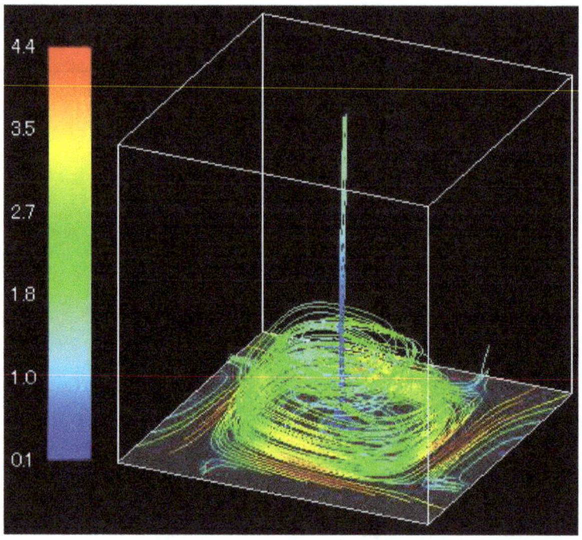

Figure 10. Streamlines of whirlwind

Figure11 shows the upward velocity distribution along the axis from the bottom center. The velocity is almost proportional to the height, and then is toward to a uniform values of about 9 m/s, which is larger than that of inducted air current. Results from the combined radiative-convective flow analysis are almost the same with that from convective analysis.

Figure12 shows the circulation (product of radial distance and circumefential velocity) distribution at the 10m height from the bottom surface. Almost the inner part of whirlwind is assumed to be a forced vortex, and the outer part is to be a free vortex, due to the four air inlets. Results from the combined radiative-convective flow analysis are also almost the same with that from convective analysis.

Figure13 shows the radial position of the maximum circulation for the height, and Figure14 shows the values of the maximum circulation for the height. Figure13 indicates again that the whirlwind is shrinked sharply with the height from the heat source, and Figure14 declares that the maximum circulation is positioned at 100m from the bottom surface due to the friction and 200m height of air current inlet. Results from the combined radiative-convective flow analysis are almost the same with that from convective analysis again.

3.2.2.3. Summary

From the comparison of thermal and flow field between convective flow analysis and combined radiative-convective flow analysis, radiative heat exchange has a great influence to the thermal field and a less influence to the flow field. Flow field is much characterised by turbulent.

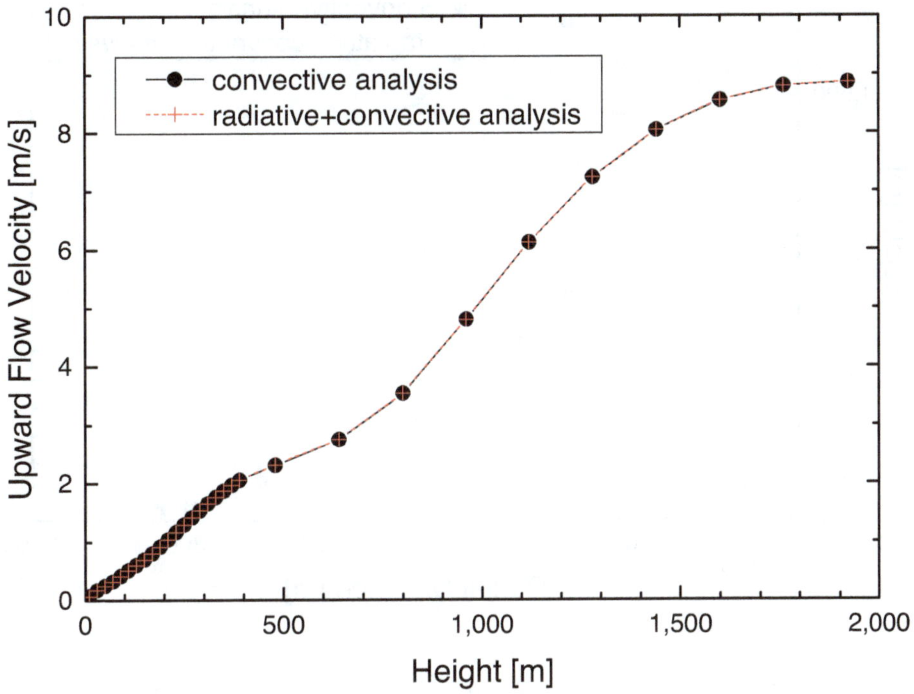

Figure 11. Upward velocity distribution along the axis from the bottom center

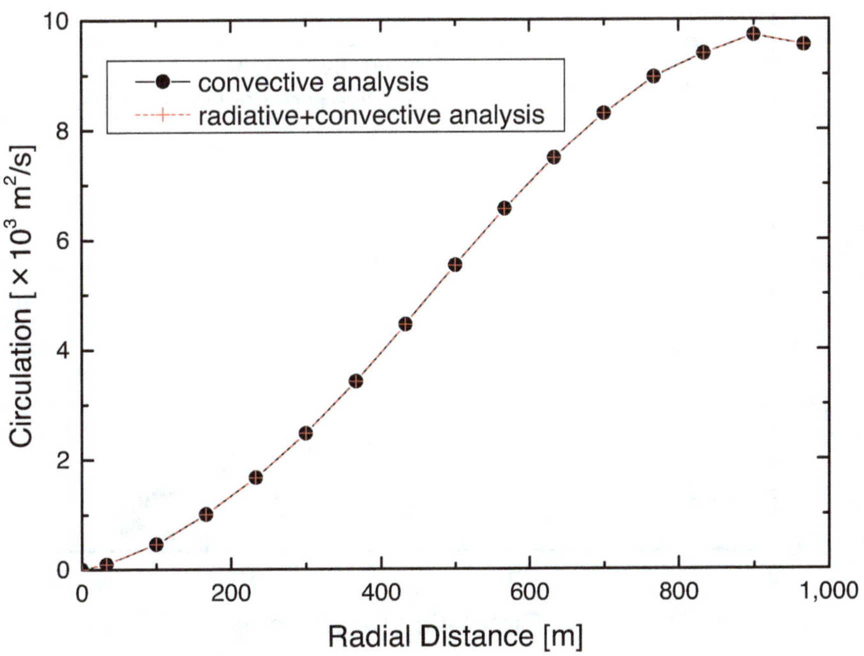

Figure 12. Circulation distribution at the 10m height from the bottom surface

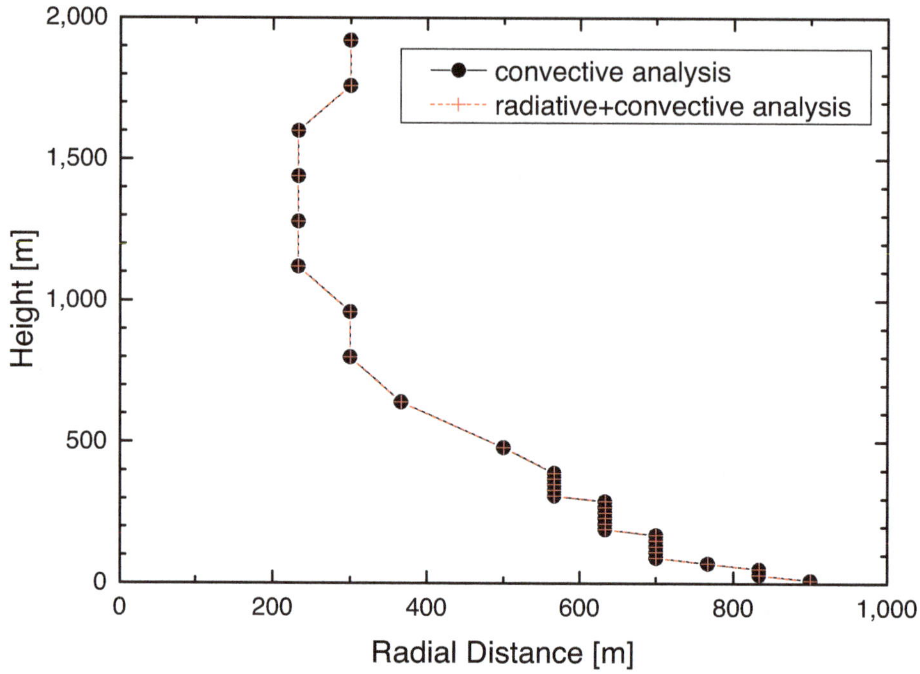

Figure 13. Radial position of the maximum circulation for the height

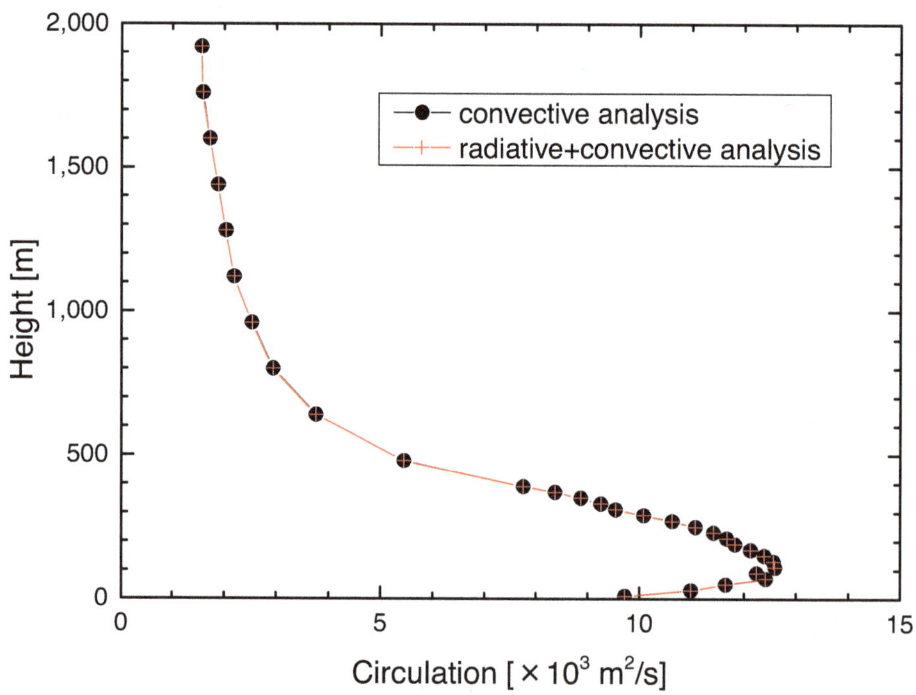

Figure 14. Maximum circulation for the height

3.2.3. *Influence of participating media concentration on heat exchange and flow distribution*

In practical fire whirlwind, combustion and chemical reaction release some participating media. In this section, representative participating media, i.e. carbon dioxide and water vapor is considered, and the concentration of these participating media is altered to evaluate the influence on heat exchange and flow distribution.

Figure15 shows a comparison of divergence of radiative heat flux with changing the concentration of participating media along the Table 1. Case A means no participating media in the fluid, resulting in no radiative heat exchange. Increase of concentration of participating gases leads the increase of divergence of radiative heat flux in the vicinity of the heat source, i.e. absorption and re-emission of heat due to these participating media. Remarkable increase of divergence of radiative heat flux is observed in case of increase of water vapor concentration.

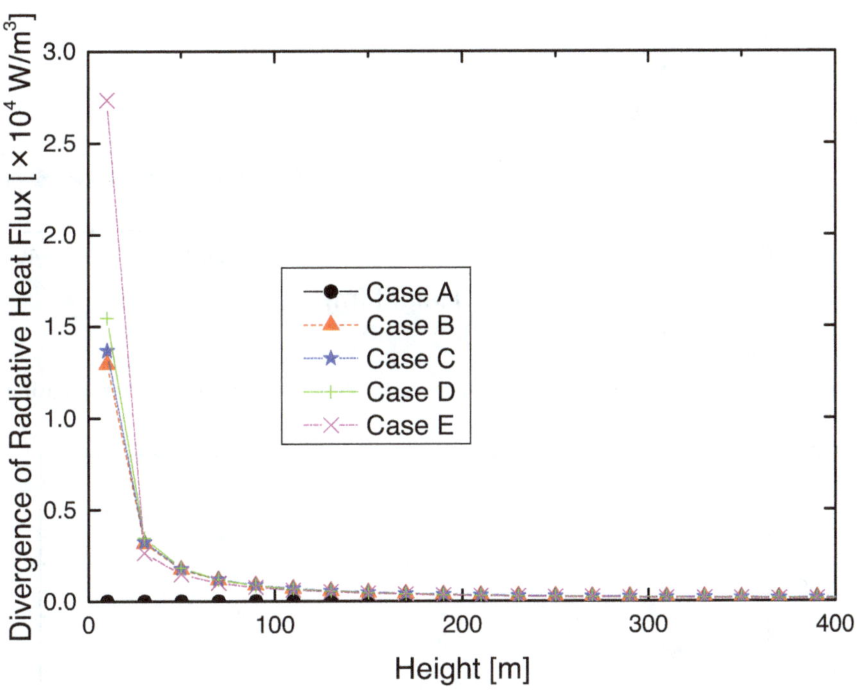

Figure 15. Comparison of divergence of radiative heat flux with changing the concentration of participating media

Figure16 shows a comparison of temperature distribution above the heat source after 30 minutes from air current induction. Increase of divergence of radiative heat flux due to the increase of participating media concentration leads raise of temperature, especially in the vicinity of the heat source.

Figures 17 and 18 shows temperature at the height of 10m from the bottom surface with changing the concentration of carbon dioxide and that of water vapor, respectively. Some values of both participating media concentration are added from the Table 1 for the combined analysis to observe the tendency between temperature and concentration. As discussed in Figure 15, water vapor plays more important role to the thermal field than carbon dioxide.

It is summarized that increase of participating media concentration gives raise of temperature due to absorption and re-emission, and water vapor influences thermal field more than carbon dioxide. However, these calculations employ uniform concentration over the analytical domain. Release and diffusion of participating media have to be considered for more practical evaluation of the whirlwind.

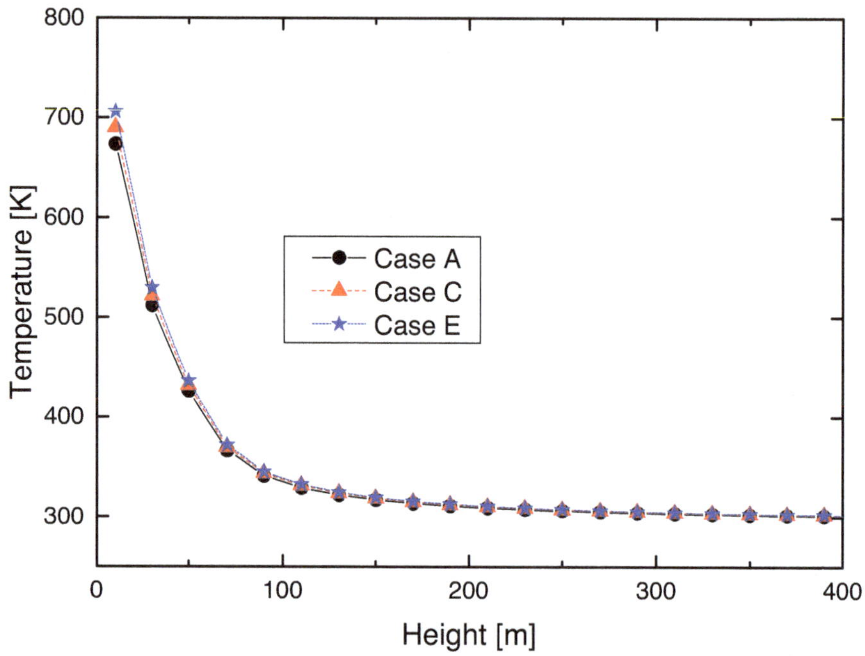

Figure 16. Comparison of temperature distribution with changing the concentration of participating media

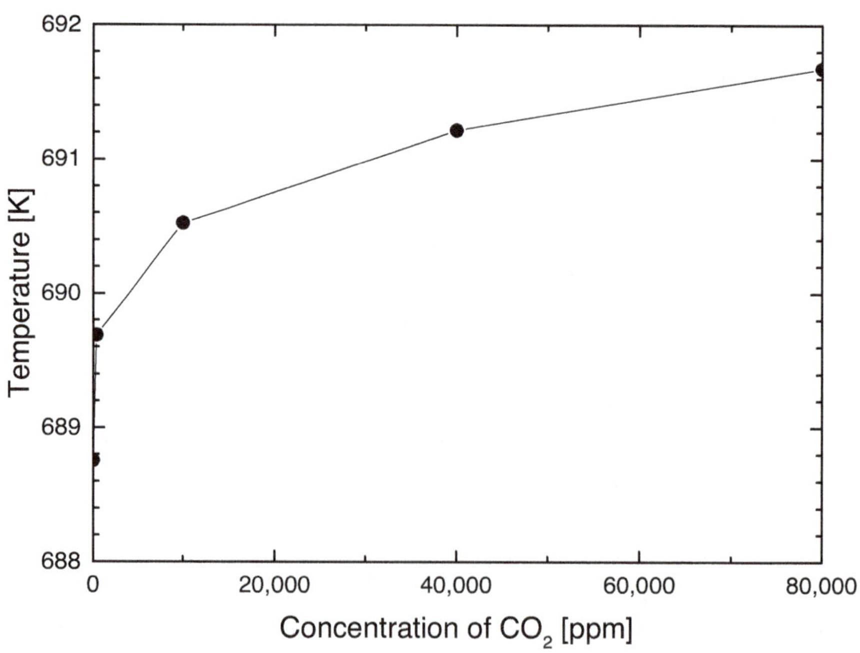

Figure 17. Temperature at the height of 10m from the bottom surface with changing the concentration of carbon dioxide

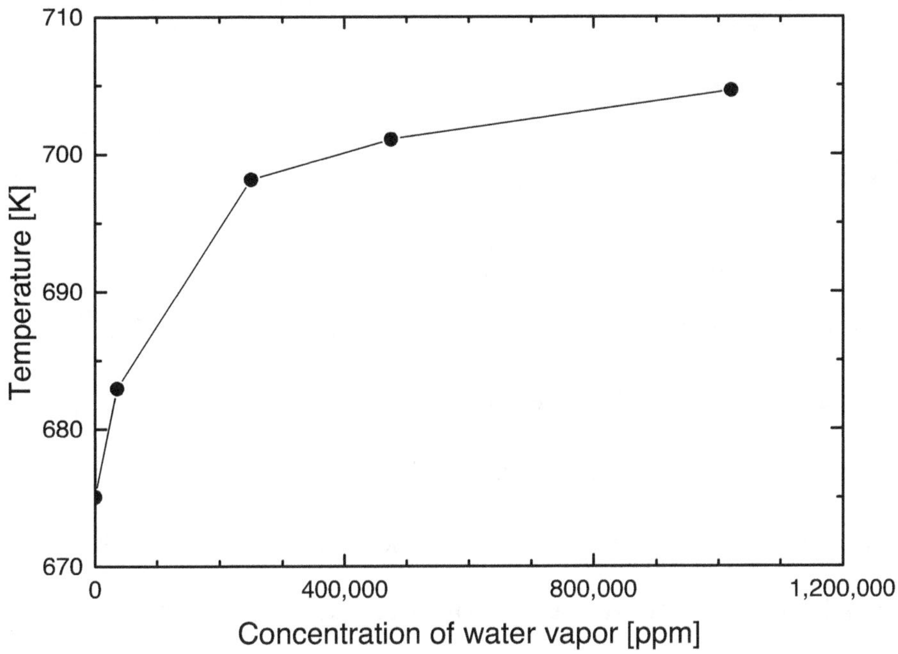

Figure 18. Temperature at the height of 10m from the bottom surface with changing the concentration of water vapor

4. Conclusions

In this study, three dimensional analyses are performed to investigate the thermal and flow fields by using the Finite Volume Method with introducing divergence of radiative heat flux for gas medium. Fire whirlwind is forcely generated stably just above the heat source with introducing air currents from four courners. One dimensional radiative exchange analysis above the heat source is compared with three dimensional one to reduce the computational load and time. One-dimensional radiative exchange analysis is suffifient with respect to the accuracy in these calculations. Then, the composition of participating gases is altered to discuss the effect of radiative heat exchange to the whirlwind flow field. The following concluding remarks are gotten from the combined heat transfer analysis.

- Individual calculation of convective flow analysis and radiative exchange suggest utilization of at least 30 non-uniform mesh spacings for z direction, and 30x30x30 mesh with non-uniform spacings is adapted for combined radiative-convective analysis.
- From the comparison of thermal and flow field between convective flow analysis and combined radiative-convective flow analysis, radiative heat exchange has a great influence to the thermal field and a less influence to the flow field. Flow field is much characterised by turbulent.
- Increase of participating media concentration gives raise of temperature due to absorption and re-emission, and water vapor influences thermal field more than carbon dioxide. However, these calculations employ uniform concentration over the analytical domain.

Release and diffusion of participating media have to be considered for more practical evaluation of the whirlwind.

Nomenclature

$A(\hat{s})$; Area projected onto the surface normal to $\hat{s}$

$A_i(\hat{s})$; Area of i-th radiation element projected onto the surface normal to $\hat{s}$

$A_{i,\lambda}^{R}$; Effective radiation area of i-th radiation element

$F_{i,j}^{A}$; Absorption view factor from i-th radiation element to j-th radiation element

$F_{i,j}^{D}$; Diffuse scattering view factor from i-th radiation element to j-th radiation element

$I_{b,i,\lambda}$; Spectral black-body radiation intensity of i-th radiation element

$I_{b,\lambda}$; Spectral black-body radiation intensity

I_{λ} ; Spectral radiation intensity

$I_{i,\lambda}^{D}$; Average scattered radiant intensity of i-th radiation element

I_{λ}^{D} ; Average scattered radiant intensity

$Q_{J,i,\lambda}$; Spectral radiation energy of i-th radiation element

$Q_{T,i,\lambda}$; Spectral heat transfer rate of the emissive power of i-th radiation element

$Q_{X,i,\lambda}$; Spectral net rate of heat generation of i-th radiation element

$q_{X,i}$; Heat generation rate of i-th radiation element per unit volume

$\vec{r}$; Position vector

$\vec{r}_0$; Position vector

S ; Path length

$\bar{S}_i(\hat{s})$; Path length of i-th radiation element ($\bar{S}_i(\hat{s})=V_i/A_i(\hat{s})$)

$\hat{s}$; Direction vector

$\hat{s}'$; Direction vector

T ; Temperature

T_i ; Temperature of i-th radiation element

t; Time

u ; Normalized velocity component for x direction

V_i ; Volume of i-th radiation element

v; Normalized velocity component for y direction

w ; Normalized velocity component for z direction

Greek symbols

$\beta_{i,\lambda}^{*}$; Apparent extinction coefficient of i-th radiation element

β_{λ}^{*} ; Apparent extinction coefficient

Γ ; Generalized diffusion coefficient

$\varepsilon_{i,\lambda}$; Spectral emissivity of i-th radiation element

κ_{λ} ; Spectral absorption coefficient

λ ; Wave length

μ' ; Directional cosine

$\sigma_{s,\lambda}$; Spectral scattering coefficient

$\Phi_{\lambda}(\hat{s}' \to \hat{s})$; Phase function from the direction $\hat{s}'$ to $\hat{s}$

ϕ ; Variable for governing equations

$\Omega_{i,\lambda}^{D}$; Spectral diffuse reflectivity of i-th radiation element

$\Omega_{i,\lambda}^{S}$; Spectral specular reflectivity of i-th radiation element

Ω_{λ} ; Scattering albedo

Ω_{λ}^{*} ; Corrected scattering albedo

ω ; Solid angle

Acknowledgements

This work was partially supported by JSPS KAKENHI Grant Number 24560222.

Author details

Seigo Sakai*

Yokohama National University, Japan

References

[1] Takewaki I, Murakami S, Fujita K, Yoshitomi S, Tsuji M. The 2011 off the Pacific coast of Tohoku earthquake and response of high-rise buildings under long-period ground motions. Soil Dynamics and Earthquake Engineering 2011; 31(11) 1511–1528.

[2] Hough S E, Bilham R G. After the Earth Quakes: Elastic Rebound on an Urban Planet, Oxford University Press USA 2005.

[3] Graham H E. Fire whirlwinds. Bulletin American Meteorological Society 1955; 36 99–103.

[4] Emmons H W, Ying S J. The fire whirl. Proceedings of the Combustion Institute 1967; 11 475–488.

[5] Byram G M, Martin R E. The Modeling of Fire Whirlwinds. Forest Science 1970; 16(4) 386-399.

[6] Haines D A, Updike G H. Fire Whirlwind Formation over Flat Terrain. USDA Forest Service Research Paper 1971; NC-71.

[7] Martin R E, Pendleton D W, Burgess W. Effect of Fire Whirlwind Formation on Solid Fuel Burning Rates. Fire Technology 1976; 12(1) 33-40.

[8] Muraszew A, Fedele J B, Kuby W C. The fire whirl phenomenon. Combustion and Flame 1979; 34 29-45.

[9] Emori R I, Saito K. Model experiment of hazardous forest fire whirl. Fire Technology 1982; 18(4) 319-327.

[10] Satoh K, Yang K T. Experimental Observations of Swirling Fires. Proceedings of ASME Heat Transfer Division 1996; HTD-335(4) 393-400.

[11] Hayashi Y, Ohmiya Y, Iwami T, Saga T. Experimental Study on Fire and Plume Properties Using BRI's Fire Wind Tunnel Facility. International Journal for Fire Science and Technology 2003; 22 17-35.

[12] Liu N A. Experimental and Theoretical Investigation on Fire Interactions and the Induced Firewhirls in Square Fire Arrays. Proceedings of Fifth NRIFD International Symposium on Forest Fires, 2005; Tokyo 293-301.

[13] Kuwana K, Sekimoto K, Saito K, Williams F A. Can We Predict the Occurrence of Extreme Fire Whirls? AIAA JOURNAL 2007; 45 16-19.

[14] Liu N A, Liu Q, Deng Z H, Satoh K, Zhu J P. Burn-out Time Data Analysis on Interaction Effects among Multiple Fires in Fire Arrays. Proceedings of the Combustion Institute 2007; 31 2589-2597.

[15] Kuwana K, Sekimoto K, Saito K, Williams F A. Scaling fire whirls. Fire Safety Journal 2008; 43(4) 252-257.

[16] Chuah K H, Kuwana K, Saito K, Williams F A. Inclined fire whirls. Proceedings of the Combuation Institute 2011; 33 2417-2424.

[17] Satoh K, Yang K T. Simulations of swirling fires controlled by channeled self-generated entrainment flows [A]. Fire Safety Sci., Proceedings of the 5th International Symposium[C] 1997; 201-212.

[18] Battaglia F, Rehm R G, Baum H R. Fluid Mechanics of Fire Whirls: An Inviscid Model. Physics of Fluids 2000; 12(11) 2859-2867.

[19] Battaglia F, McGrattan K B, Rehm R G, Baum H R. Simulating Fire Whirls. Combustion Theory and Modelling 2000; 4(3) 123-138.

[20] Snegirev A Y, Marsden J A, Francis J, Makhviladze G M. Numerical studies and experimental observations of whirling flames. International Journal of Heat and Mass Transfer 2004; 47 2523–2539.

[21] Hassan M I, Kuwana K, Saito K, Wang F. Flow Structure Of A Fixed-frame Type Firewhirl. Proceedings of the Eighth International Symposium, International Association for Fire Safety Science, 18-23 September 2005, Beijing, China. 2005; 951-962.

[22] Chuah K H, Kushida G. The prediction of flame heights and flame shapes of small fire whirls. Proceedings of the Combustion Institute 2007; 31 2599–2606.

[23] Grishin A M. Effect of the Interaction between Fire Tornadoes on Their Propagation. Doklady Physics 2007; 52(10) 521-522.

[24] Grishin A M, Matvienko O V, Rudi Y A. Mathematical Modeling of Gas Combustion in A Twisted Jet and of The Formation of A Fiery Whirlwind. Journal of Engineering Physics and Thermophysics 2009; 82(5) 906-913.

[25] Sakai S, Watanabe Y. Numerical Study of Interaction between Natural Convection Flow and horizontal wind. Proceedings of the FEDSM2007, 5th Joint ASME/JSME Fluids Engineering Conference, 30 July – 2 August 2007, San Diego, California, USA. 2007; FEDSM2007-37212.

[26] Sakai S, Miyagi N. Numerical Study of Fire Whirlwind Taking into Account Radiative Heat Transfer, IOP Conference Series: Materials Science and Engineering 2010; 10 012031.

[27] Sakai S. Numerical Prediction of Fire Whirlwind Outbreak and Scale Effect of Whirlwind Behavior. In: Moustafa A. (ed.) Advances in Geotechnical Earthquake Engineering – Soil Liquefaction and Seismic Safety of Dams and Monuments. Rijeka: In Tech; 2012. p383-404.

[28] Maruyama S, Aihara T. Radiation Heat Transfer of Arbitrary Three-Dimensional Absorbing, Emitting and Scattering Media and Specular and Diffuse Surfaces. Trans. ASME, Journal of Heat Transfer 1997; 119 129-136.

[29] Patankar S V. Numerical heat transfer and fluid flow. In: Phillips M A, Millman E M. (ed.) USA: Hemisphere Publishing Corporation; 1980.

[30] Maruyama S, Mori Y, Sakai S. Nongray radiative heat transfer analysis in the anisotropic scattering fog layer subjected to solar irradiation. Journal of Quantitative Spectroscopy and Radiative Transfer 2004; 83 361-375.

[31] Maruyama S. Radiative heat transfer in anisotropic scattering media with specular boundary subjected to collimated irradiation. International Journal of Heat and Mass Transfer 1998; 41 2847-2856.

[32] Nishikawa T, Maruyama S, Sakai S. Radiative Heat Transfer Analysis within Three-Dimensional Clouds Subjected to Solar and Sky Irradiation. Journal of the Atmospheric Sciences 2004; 61 3125-3133.

4

Initial Shapes of Cable-Stayed Bridges during Construction by Cantilever Methods – Numerical Simulation and Validation of the Kao Ping Hsi Bridge

Ming-Yi Liu and Pao-Hsii Wang

1. Introduction

Cable-stayed bridges have become popular because of their aesthetic appearance, structural efficiency, ease of construction and economic advantage over the past several decades. However, this kind of bridge is light, flexible and with low inherent damping. Accordingly, they are sensitive to ambient excitations from seismic, wind and traffic loads. Because the geometric and dynamic properties of the bridges as well as the characteristics of the excitations are complicated, it is important to fully understand the system behaviors with reasonable bridge shapes at each erection stage during construction by the cantilever method, which is used to provide the necessary information to accurately calculate the dynamic responses of the bridges under the complex excitations.

A lot of studies on this kind of bridges have been done in the last half century [1-3]. However, few analytical techniques have been presented for the cable-stayed bridges during erection stages in construction. To the authors' best knowledge, a number of papers have investigated the erection procedure of cable-stayed bridges, focusing primarily on the improvement of construction technology [4-6], but not for the analytical purpose. Since the initial shape of a cable-stayed bridge not only reasonably provides the geometric configuration as well as the prestress distribution of the whole bridge under the weight of the deck-tower system and the pretension forces in the stay cables, but also definitely ensures the satisfaction of the relationships for the equilibrium conditions, boundary conditions and architectural design requirements [7-11], it is essential to research the system behaviors with the appropriate initial shapes of cable-stayed bridges.

The objective of this chapter is to fully understand the system behaviors with the appropriate initial shapes of cable-stayed bridges at each erection stage during construction by the cantilever method. Two computational procedures during erection stages: a forward process analysis and a backward process analysis, are presented for this reason [10]. On the basis of the two procedures, a series of initial shape analyses are conducted to investigate the bridge shapes at each erection stage. Numerical examples based on finite element models of the Kao Ping Hsi Bridge in Taiwan [12] are developed to validate the two proposed approaches. The initial shape at each erection stage provides the necessary data for checking and controlling the real-time erection procedure of a cable-stayed bridge during construction. The designed shape, i.e., the geometric configuration and the prestress distribution, of the whole bridge can then be achieved.

2. Finite element formulation

A cable-stayed bridge can be considered as an assembly of a finite number of cable elements for the stay cables and beam-column elements for both the decks and towers based on the finite element concepts. A number of assumptions are used in this study: the material is homogeneous and isotropic; the stress-strain relationship of the material remains within the linear elastic range during the whole nonlinear response; the external forces are displacement independent; large displacements and large rotations are allowed, but strains are small; each stay cable is fixed to both the deck and tower at their joints of attachment. Under the above assumptions, the initial shape analysis of cable-stayed bridges is conducted according to the system equations with the consideration of geometric nonlinearities.

2.1. Geometric nonlinearities

Three types of geometric nonlinearities: the cable sag, beam-column and large displacement effects, are considered to reasonably simulate cable-stayed bridges in this study.

A stay cable will sag into a catenary shape because of its weight and tensile force. Such cable sag effect has to be considered when the stay cable is represented by a single straight cable element. A stay cable with tensile stiffness is assumed to be perfectly elastic. The compressive, shear and bending stiffnesses of the stay cable are neglected. The cable sag nonlinearity can be simulated according to the equivalent modulus of elasticity of the stay cable [13]

$$E_{eq} = \frac{E_c}{1 + \frac{(wl_c)^2 A_c E_c}{12T^3}}, \tag{1}$$

where E_c, A_c and l_c are the effective modulus of elasticity, the cross-sectional area and the horizontal projected length of the stay cable, respectively; w is the weight of the stay cable per unit length; T is the tension in the stay cable. The stiffness matrix of a cable element in Figure 1 can be expressed as

$$KE_{jk} = \begin{cases} \left[\dfrac{E_{eq}A_c}{L_c}\right], & u_1 > 0 \\ [0], & u_1 \leq 0 \end{cases}, \tag{2}$$

where u_1 is the element coordinate for the relative axial deformation; L_c is the chord length of the stay cable.

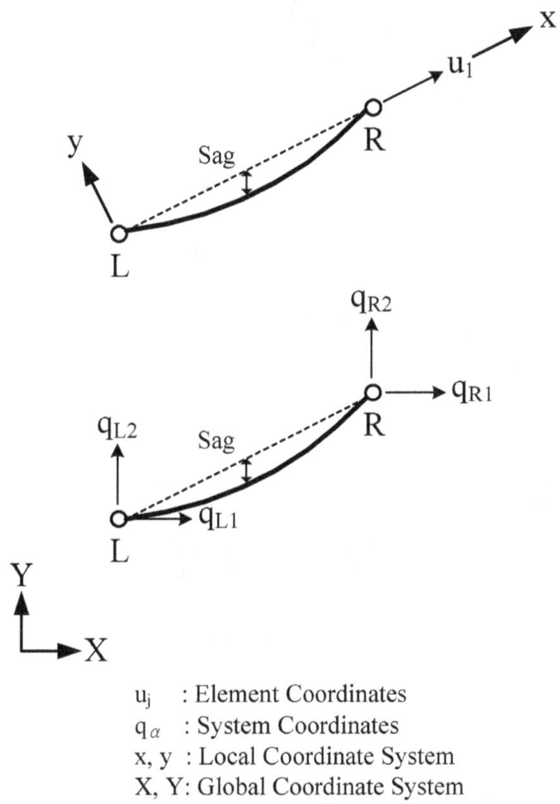

Figure 1. Cable element.

Large compressive forces in the deck-tower system of a cable-stayed bridge occur due to high pretension forces in the stay cables. For this reason, the beam-column effect between such compressive forces and bending moments has to be taken into consideration when beam-column elements are used to simulate both the decks and towers. Shear strains of a beam-column element in Figure 2 are negligible according to the Euler-Bernoulli beam theory. u_1, u_2 and u_3 are the element coordinates for the left end rotation, the right end rotation and the relative axial deformation, respectively. The stiffness matrix of the beam-column element can be written as

$$KE_{jk} = \frac{E_b I_b}{L_b}\begin{bmatrix} C_s & C_t & 0 \\ C_t & C_s & 0 \\ 0 & 0 & R_t A_b / I_b \end{bmatrix}, \tag{3}$$

where E_b, A_b, I_b and L_b are the modulus of elasticity, the cross-sectional area, the moment of inertia and the length of the beam-column element, respectively; C_s, C_t and R_t are the stability functions representing the interaction between the axial and bending stiffnesses of the beam-column element [14].

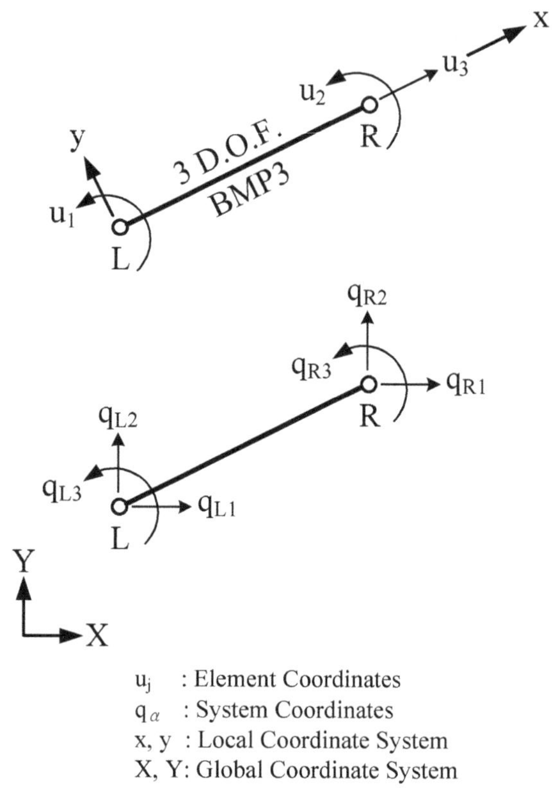

Figure 2. Beam-column element.

Large displacements occur in the deck-tower system due to the large span as well as less weight of a cable-stayed bridge. Such effect has to be considered when the equilibrium equations are derived from the deformed position. Under these conditions, the element coordinate u_j can be expressed as a nonlinear function of the system coordinate q_α in both Figure 1 and Figure 2, i.e., $u_j = u_j(q_\alpha)$. By differentiating u_j with respect to q_α, the first-order and second-order coordinate transformation coefficients can be individually written as

$$a_{j\alpha} = \frac{\partial u_j}{\partial q_\alpha}, \tag{4}$$

$$a_{j\alpha,\beta} = \frac{\partial a_{j\alpha}}{\partial q_\beta} = \frac{\partial^2 u_j}{\partial q_\alpha \partial q_\beta}. \tag{5}$$

$a_{j\alpha}$ and $a_{j\alpha,\beta}$ for the stiffness matrices of the cable and beam-column elements can be found in [7], which are used to develop the tangent system stiffness matrix in Section 2.2.

2.2. System equations

The system equations in generalized coordinates of a nonlinear finite element model of a cable-stayed bridge can be derived from the Lagrange's virtual work principle

$$\sum_{EL} S_j a_{j\alpha} = P_\alpha, \alpha = 1,2,3,\cdots,N, \tag{6}$$

$$S_j = KE_{jk} u_k + S_j^0, \tag{7}$$

$$P_\alpha = \vec{K}^j \cdot \vec{b}_\alpha^j, \tag{8}$$

$$\vec{b}_\alpha^j = \frac{\partial \vec{W}^j}{\partial q_\alpha}, \tag{9}$$

where S_j is the element force vector; P_α is the external force vector; S_j^0 is the initial element force vector; $\vec{K}^j$ is the external nodal force vector; $\vec{b}_\alpha^j$ is the basis vector; $\vec{W}^j$ is the displacement vector corresponding to $\vec{K}^j$; N is the number of degrees of freedom; the subscript α denotes the number of the system coordinate; the subscripts j and k represent the numbers of the element coordinates; the superscript j denotes the nodal number; $\sum_{EL}$ represents the summation over all elements.

Under consideration of three types of geometric nonlinearities mentioned in Section 2.1, KE_{jk} of a cable element and that of a beam-column element can be determined from Eq. (2) and Eq. (3), respectively. The former and the latter are individually due to the cable sag effect and the beam-column effect. u_j, $a_{j\alpha}$ and $\vec{b}_\alpha^j$ are nonlinear functions of q_α when the large displacement effect occurs. $\vec{K}^j$ can be written as a function of q_α if they are displacement dependent forces.

Eq. (6) is a set of simultaneous nonlinear equations. In order to incrementally solve these equations, the linearized system equations in a small force interval are derived based on the first-order Taylor series expansion of Eq. (6)

$${}^2K_{\alpha\beta}^n \Delta q_\beta^n = {}_uP_\alpha^n + \Delta P_\alpha^n, \; P_\alpha^n \le P_\alpha \le P_\alpha^{n+1}, \tag{10}$$

$${}^2K_{\alpha\beta}^n = \sum_{EL} KE_{jk}^n a_{j\alpha}^n a_{k\beta}^n + \sum_{EL} S_j^n a_{j\alpha,\beta}^n - {}^n\vec{K}^j \cdot {}^n\vec{b}_{\alpha,\beta}^j - {}^n\vec{K}_\beta^j \cdot {}^n\vec{b}_\alpha^j, \tag{11}$$

$$\vec{b}_{\alpha,\beta}^j = \frac{\partial \vec{b}_\alpha^j}{\partial q_\beta}, \tag{12}$$

$$\vec{K}_\alpha^j = \frac{\partial \vec{K}^j}{\partial q_\alpha}, \tag{13}$$

$$_{u}P_{\alpha}^{n} = P_{\alpha}^{n} - \sum_{EL} S_{j}^{n} a_{j\alpha}^{n}, \tag{14}$$

$$\Delta P_{\alpha}^{n} = P_{\alpha}^{n+1} - P_{\alpha}^{n}, \tag{15}$$

$$\Delta q_{\alpha}^{n} = q_{\alpha}^{n+1} - q_{\alpha}^{n}, \tag{16}$$

where ${}^{2}K_{\alpha\beta}^{n}$ is the tangent system stiffness matrix; ${}_{u}P_{\alpha}^{n}$ is the unbalanced force vector; ΔP_{α}^{n} is the increment of the external force vector; Δq_{α}^{n} is the increment of the system coordinate vector; the superscripts n and $n+1$ denote the numbers of the force steps; the superscript 2 represents the second-order iteration matrix.

${}^{2}K_{\alpha\beta}^{n}$ in Eq. (11) includes four terms. The first term is the elastic stiffness matrix, while the second and third terms are the geometric stiffness matrices induced by large displacements. In addition, the fourth term is the geometric stiffness matrix induced by displacement dependent forces, which is negligible in this study.

Eq. (10) is a set of simultaneous linear equations in a small force interval, which can be solved by the Newton-Raphson method [7-11].

2.3. Initial shape analysis

The initial shape of a cable-stayed bridge provides the geometric configuration as well as the prestress distribution of such bridge under the weight of the deck-tower system and the pretension forces in the stay cables. The relationships for the equilibrium conditions, boundary conditions and architectural design requirements have to be achieved for the bridge shape. Under these conditions, the initial shape analysis of cable-stayed bridges is presented by considering three types of geometric nonlinearities including the cable sag, beam-column and large displacement effects.

For the initial shape analysis of a cable-stayed bridge, the weight of the deck-tower system is considered, whereas the weight of the stay cables is assumed to be negligible. The shape finding computation is conducted using a two-loop iteration method: an equilibrium iteration and a shape iteration [7-11]. It can be started with an estimated initial element force, i.e., pretension force, in the stay cables. By incrementally solving Eq. (10), i.e., the equilibrium iteration, the equilibrium configuration of the whole bridge under the weight of the deck-tower system can be obtained based on the reference configuration, i.e., architectural design form, with no deflection and zero prestress in the deck-tower system.

The bridge configuration satisfies the equilibrium and boundary conditions after the above equilibrium iteration. However, the architectural design requirements are generally not achieved due to the fact that large displacements and variable bending moments occur in the deck-tower system from the large bridge span. Under these conditions, the appropriate initial shape can be determined by conducting the shape iteration to reduce the displacements as well as to smooth the bending moments.

Several control points are selected for insuring that both the deck and tower displacements achieve the architectural design requirements in the shape iteration

$$\left|\frac{q_\alpha}{L_r}\right| \leq \varepsilon_r, \tag{17}$$

where q_α is the displacement in a certain direction of the control point; L_r is the reference length; ε_r is the convergence tolerance. Each control point is the node intersected by the deck and the stay cable for checking the deck displacement. q_α and L_r represent the vertical displacement of the control point and the main span length, respectively. Similarly, each node intersected by the tower and the stay cable, or located on the top of the tower is selected as the control point for checking the tower displacement. q_α and L_r individually denote the horizontal displacement of the control point and the tower height.

If Eq. (17) is not satisfied, the element axial forces calculated in the previous equilibrium iteration will be used as the initial element forces in the new equilibrium iteration, and the corresponding equilibrium configuration of the whole bridge under the weight of the deck-tower system will be obtained again. The shape iteration will then be repeated until Eq. (17) is satisfied. Under these conditions, the convergent configuration can be regarded as the initial shape of the cable-stayed bridge.

3. Initial shape analysis during erection stages

The initial shapes, i.e., the geometric configuration and the prestress distribution, of a cable-stayed bridge during erection stages provide the essential information for the bridge construction. According to the cantilever method, two finite element computational procedures: the forward process analysis and the backward process analysis, are presented for the shape finding of the bridge at each erection stage during construction.

3.1. Cantilever methods

The cantilever method has been widely used for the girder erection of cable-stayed bridges with self-anchored cable systems. Such method provides the natural and logical solution for the construction of large-span cable-stayed bridges during erection stages. New girder segments are installed and then supported by new stay cables at each erection stage, and the process keeps going stage-by-stage until the bridge construction is completed. The cantilever method can be further categorized into the single cantilever method and the double cantilever method. In the former, the side span girders are erected on temporary supports, while the main span girders are erected by one-sided free cantilevering until the span center or the anchor pier on the far end is reached. In the latter, the bridge girders are erected by double-sided free cantilevering from both sides of the tower toward the main span center and the anchor pier. Figure 3 illustrates the erection stages of cable-stayed bridges by the single cantilever method, which is considered in this study. Its concept can also be applied to the double cantilever method.

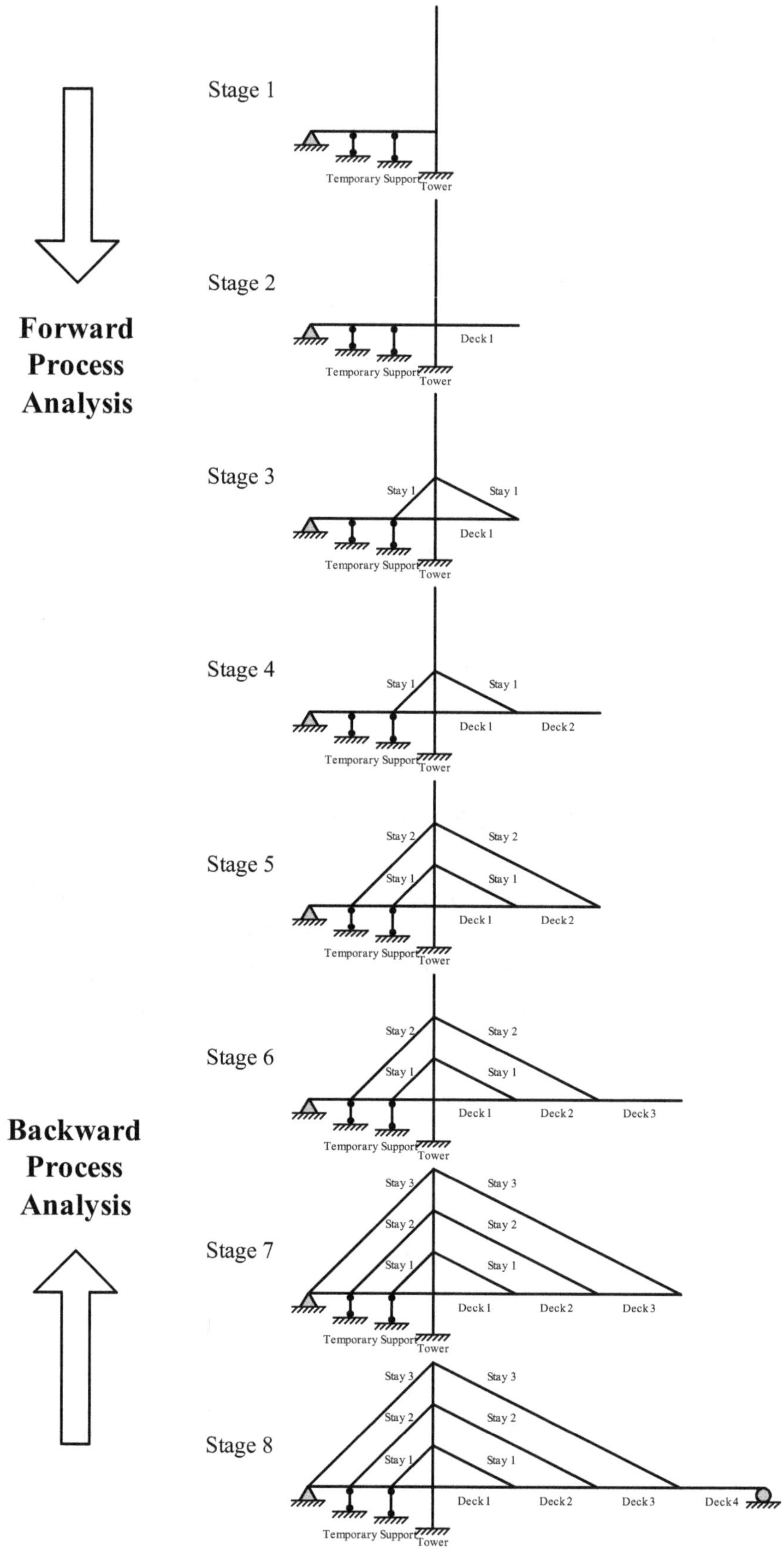

Figure 3. Erection stages of cable-stayed bridges.

3.2. Forward process analysis

The forward process analysis of cable-stayed bridges during construction by the single cantilever method is conducted based on the sequence of erection stages in Figure 3. The geometric configuration and the element forces of a cable-stayed bridge at each erection stage can therefore be estimated. At erection stages with even number (2, 4, 6), the new girder segments are installed and the corresponding relatively large vertical displacements and bending moments of the girders occur due to the lack of the exterior stay cables. While at erection stages with odd number (3, 5, 7), the new exterior stay cables are installed at the tip of the new girder segments. Under these conditions, the stay cables are stressed to lift the girders to a certain elevation, which can keep the desired correct position as well as reduce the bending moments of the girders. The pretension forces in stay cables and the girder elevation at each erection stage can be obtained by the initial shape analysis presented in Section 2.3. For the forward process analysis, the shape iteration has to be performed to keep the girders in a horizontal position, implying that an upward precamber is allowed during construction. As illustrated in Figure 4, the shape iteration can be started with an estimated initial cable force $\overset{\circ}{T}$ by setting the tip displacement of the girder segment under its dead load w and the weight of the machine equipments W_{eq} to be equal to that resulting from $\overset{\circ}{T}$, which can be written as

$$\overset{\circ}{T} = \frac{3wl + 8W_{eq}}{8\sin\alpha}, \tag{18}$$

where l is the length of the girder segment; α is the inclined angle of the stay cable.

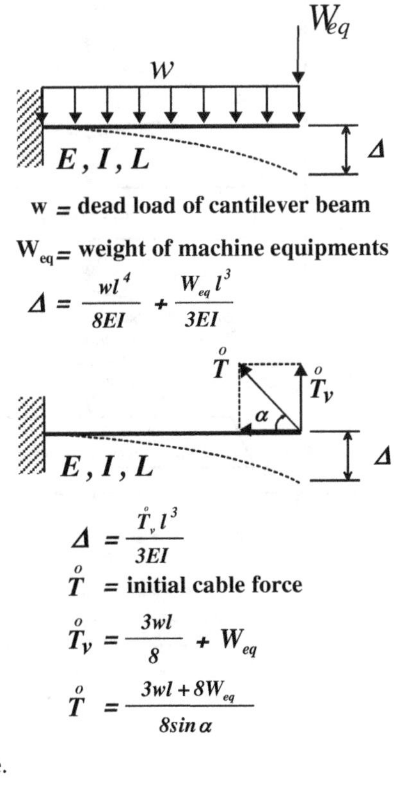

Figure 4. Estimation of initial cable force.

Figure 5 illustrates the flowchart of the forward process analysis following the actual sequence of erection stages of cable-stayed bridges during construction. The advantage of the forward process analysis is that the real-time factors of the bridges, such as creep and shrinkage of concrete, any alteration in design, etc., can be taken into consideration during construction.

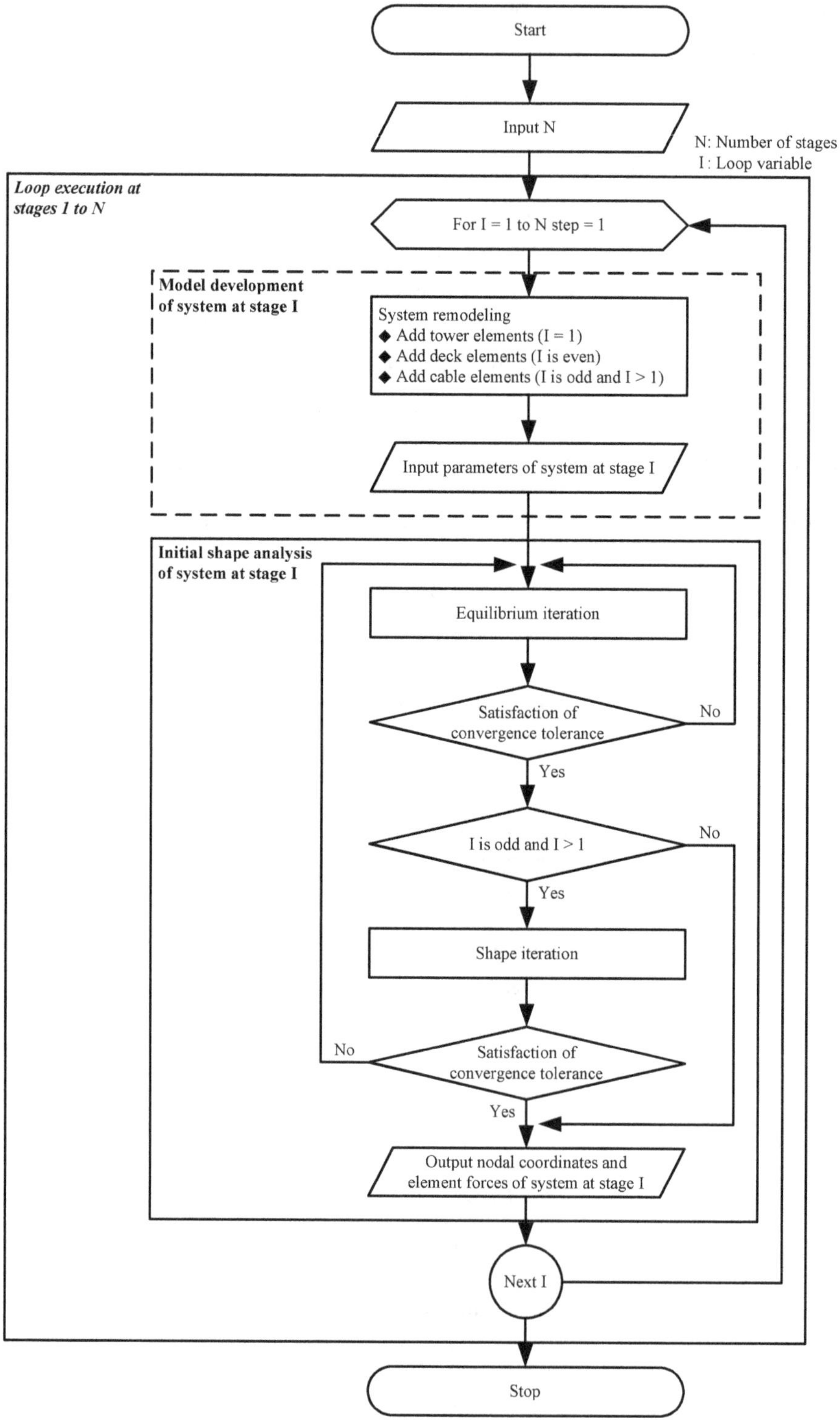

Figure 5. Flowchart of forward process analysis.

3.3. Backward process analysis

In contrast to the forward process analysis, the backward process analysis of cable-stayed bridges during construction by the single cantilever method is conducted based on the reverse sequence of erection stages in Figure 3. The computation is started with the whole bridge at the first erection stage, i.e., stage 8 in Figure 3. After removing the existing girder segments and the adjoining stay cables, the system can be remodeled and reanalyzed to estimate the geometric configuration and the member forces of the bridge at the current erection stages, i.e., stages 7 to 2 in Figure 3. The computation is continued repeatedly until the final erection stage, i.e., stage 1 in Figure 3. For the backward process analysis, the initial shape of the whole bridge has to be obtained first at the first erection stage by the initial shape analysis presented in Section 2.3. After removing the existing girder segments and the adjoining stay cables, the geometric configuration and the member forces of the bridge at each erection stage can be determined anew by solving the static system equations. On the basis of the linearized system equations from the nonlinear theory, the equilibrium iteration is performed using the Newton-Raphson method.

Figure 6 illustrates the flowchart of the backward process analysis following the reverse sequence of erection stages of cable-stayed bridges during construction. The advantage of the backward process analysis is that the initial shape of a cable-stayed bridge at each erection stage can be obtained by the equilibrium iteration without the shape iteration. Such initial shape determined by the equilibrium conditions can be considered as the desired correct position of the bridge for the next erection stage, in which the girder is precambered upwards. In other words, the computational efficiency of the backward process analysis without the shape iteration is better than that of the forward process analysis with the shape iteration. In contrast, the disadvantage of the backward process analysis is that the real-time factors of the bridge, such as creep and shrinkage of concrete, any alteration in design, etc., can not be taken into consideration during construction due to the fact that the computation is performed backwards from the whole bridge.

4. Numerical examples

As an example, the Kao Ping Hsi Bridge in Figure 7 is taken for the shape finding analysis of the bridge during the erection procedure using the single cantilever method. This bridge is an unsymmetrical single-deck cable-stayed bridge with a main span of 330 m and a side span of 184 m. The deck consisting of steel box girders in the main span and concrete box girders in the side span is supported by 28 stay cables arranged in a central plane originated at the 184 m tall, inverted Y-shaped, concrete tower. More detailed information of the Kao Ping Hsi Bridge is available in [12].

Figure 7 illustrates the two-dimensional finite element model of the bridge. This model contains 48 beam-column elements for the deck and tower, 28 cable elements for the stay cables and 49 nodes. A hinge, roller and fixed supports are used to model the boundary conditions of the left and right ends of the deck and the tower, respectively, and a rigid joint is employed to simulate the deck-tower connection. The Kao Ping Hsi Bridge was erected by single

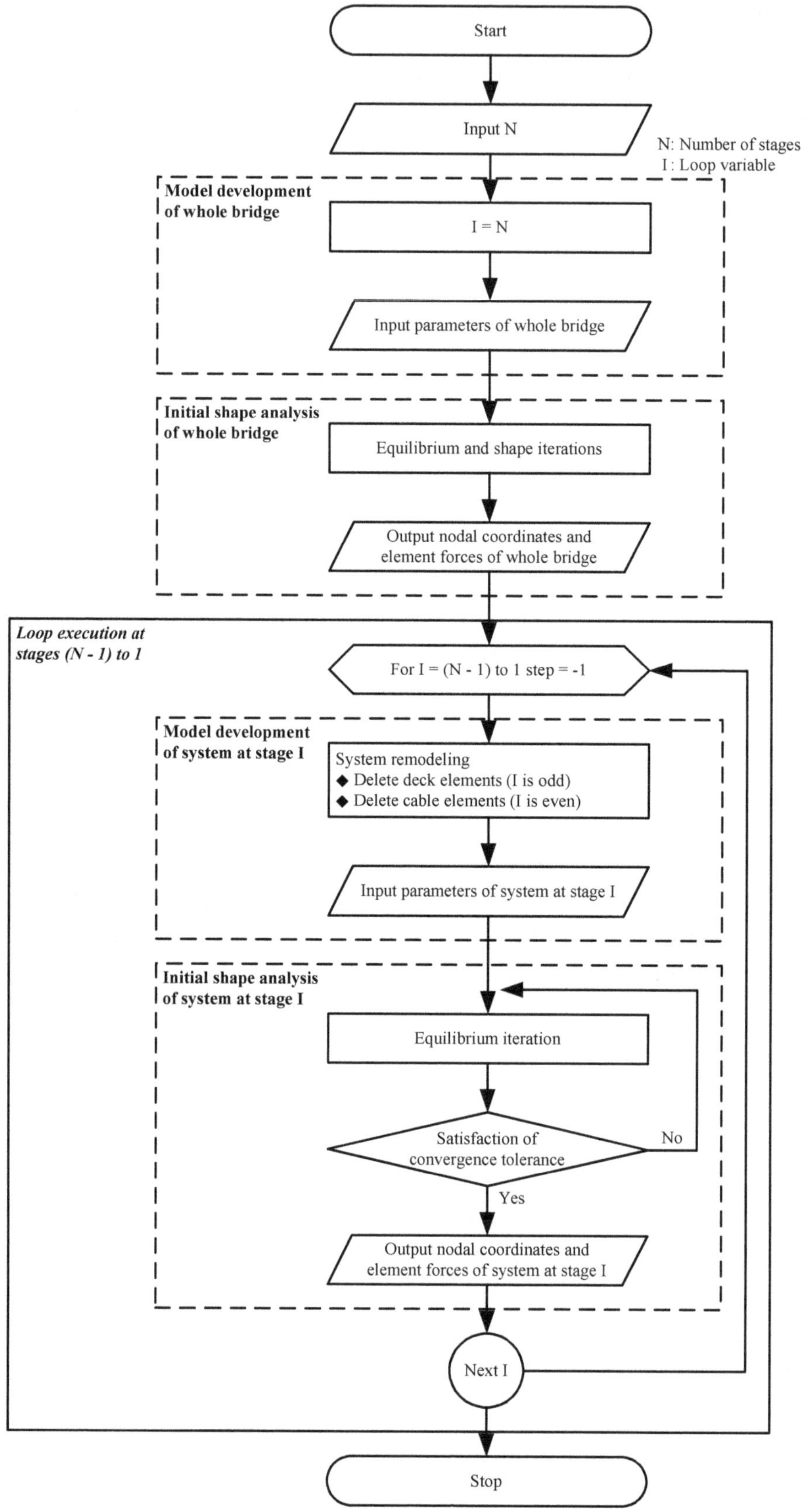

Figure 6. Flowchart of backward process analysis.

cantilever method and there are 30 erection stages for constructing the girder of the bridge. On the basis of the computational procedures during erection stages presented in Section 3, the whole bridge analysis, forward process analysis and backward process analysis are conducted in this study.

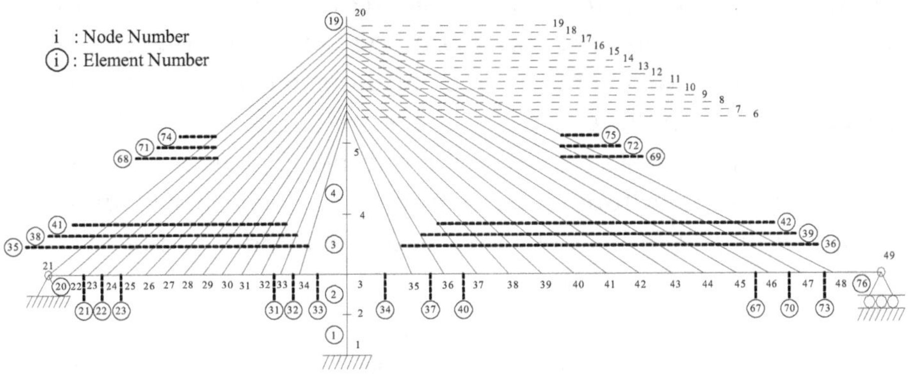

Figure 7. Finite element model of the Kao Ping Hsi Bridge.

4.1. Whole bridge analysis

Based on the finite element procedures presented in Section 2.3, the initial shape analysis is conducted to reasonably provide the geometric configuration of the whole Kao Ping Hsi Bridge. In Figure 7, nodes 27, 35, 42 and 45 are selected as the control points for checking the deck vertical displacement, while node 19 is selected as the control point for checking the tower horizontal displacement. The convergence tolerance ε_r is set to 10^{-4} in this study.

Figure 8 shows the initial shape of the whole Kao Ping Hsi Bridge (solid line), which indicates that the maximum vertical and horizontal displacements measured from the reference configuration (dashed line) are 0.033 m at node 42 in the main span of the deck and -0.024 m at node 8 in the tower, respectively. In addition, Figure 8 illustrates that the overall displacement obtained by the two-loop iteration method, i.e., the equilibrium and shape iterations, is relatively smaller than that during the shape iteration (dot-dashed line). Consequently, the initial shape obtained by the two-loop iteration method can be used to reasonably describe the geometric configurations of cable-stayed bridges.

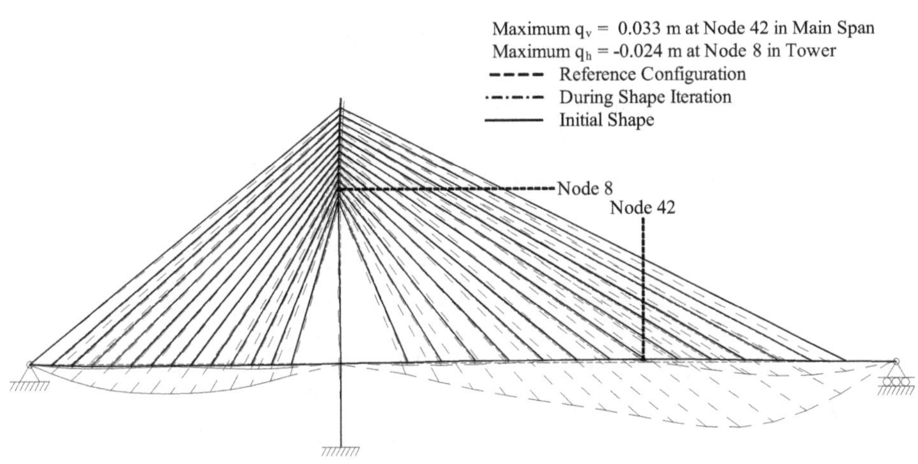

Figure 8. Initial shape of the whole Kao Ping Hsi Bridge.

4.2. Forward process analysis

According to the forward process analysis of cable-stayed bridges during construction by the single cantilever method in Section 3.2, the initial shape of the Kao Ping Hsi Bridge at each erection stage for a total of 30 stages is obtained, as shown in Figure 9. The geometric configuration of the bridge and the corresponding vertical displacement of the girder tip in the main span as well as the element forces of the exterior stay cables are illustrated at each erection stage, where NSI represents the number of shape iterations (SI) for the convergent solution. At erection stages with even number (2, 4, 6,...), the new girder segments are installed. Under these conditions, the equilibrium position can be determined anew without the shape iteration. While at erection stages with odd number (3, 5, 7,...), the new

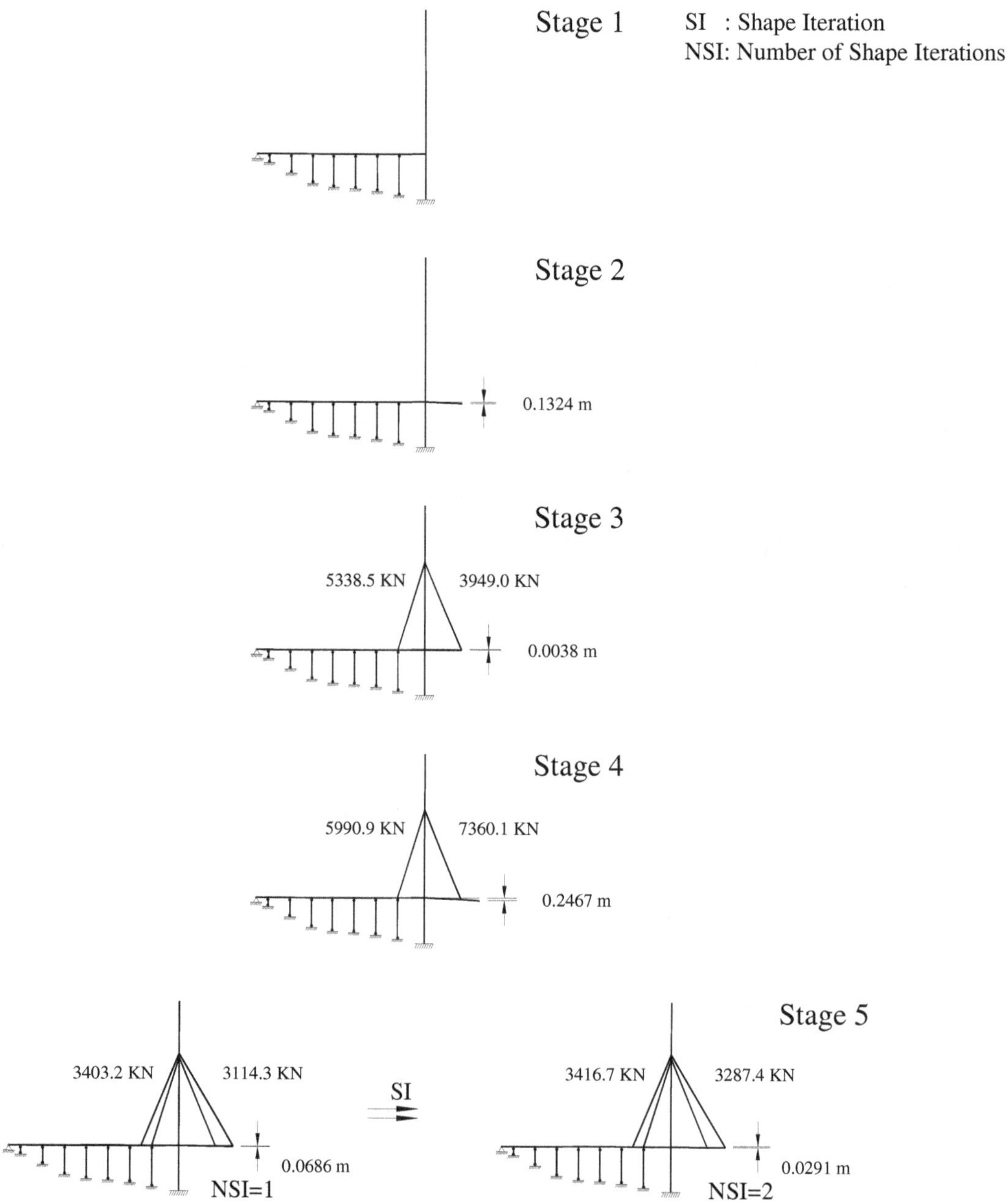

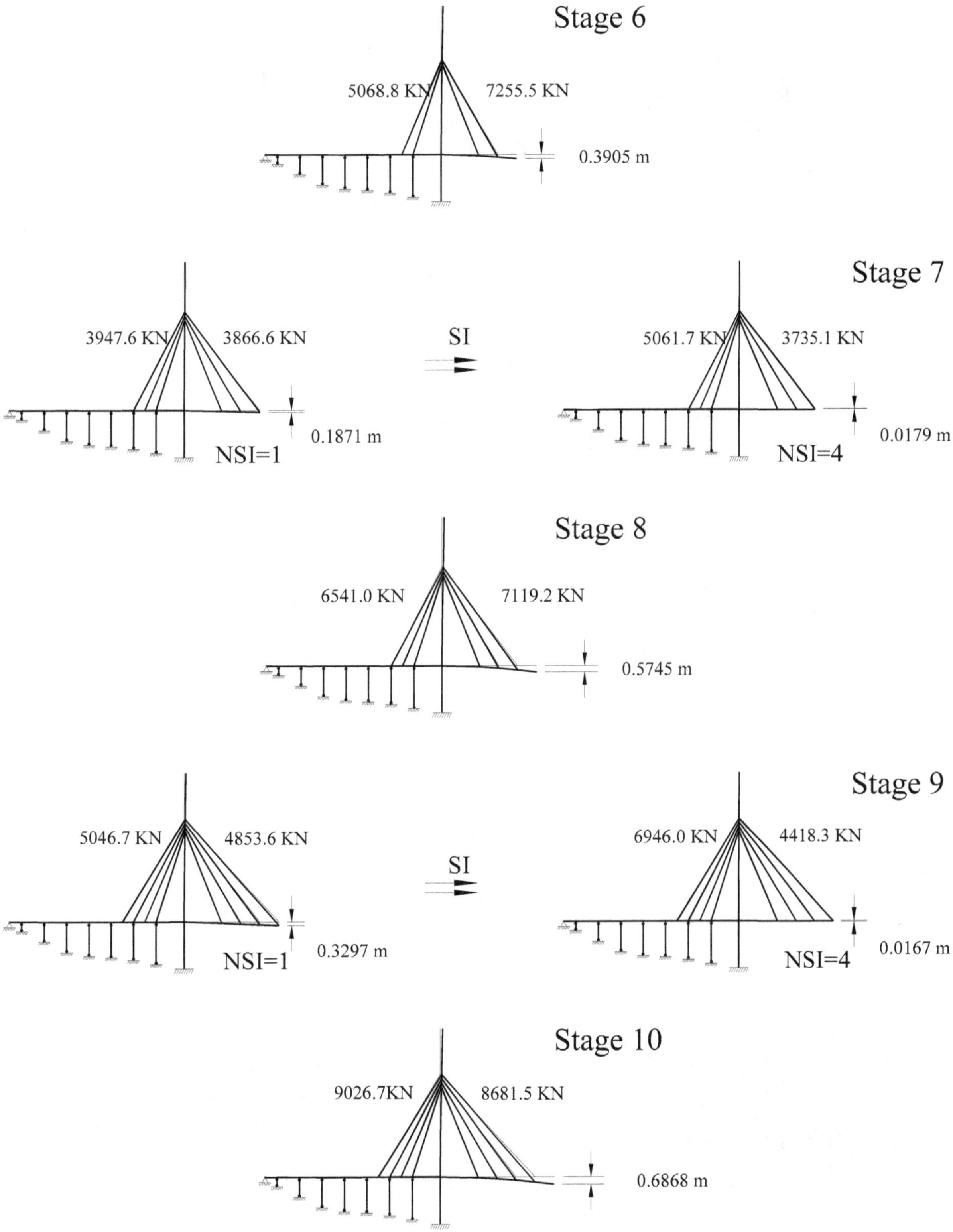
Stage 6
5068.8 KN
7255.5 KN
0.3905 m
Stage 7
3947.6 KN
3866.6 KN
0.1871 m
NSI=1
SI
5061.7 KN
3735.1 KN
0.0179 m
NSI=4
Stage 8
6541.0 KN
7119.2 KN
0.5745 m
Stage 9
5046.7 KN
4853.6 KN
0.3297 m
NSI=1
SI
6946.0 KN
4418.3 KN
0.0167 m
NSI=4
Stage 10
9026.7KN
8681.5 KN
0.6868 m

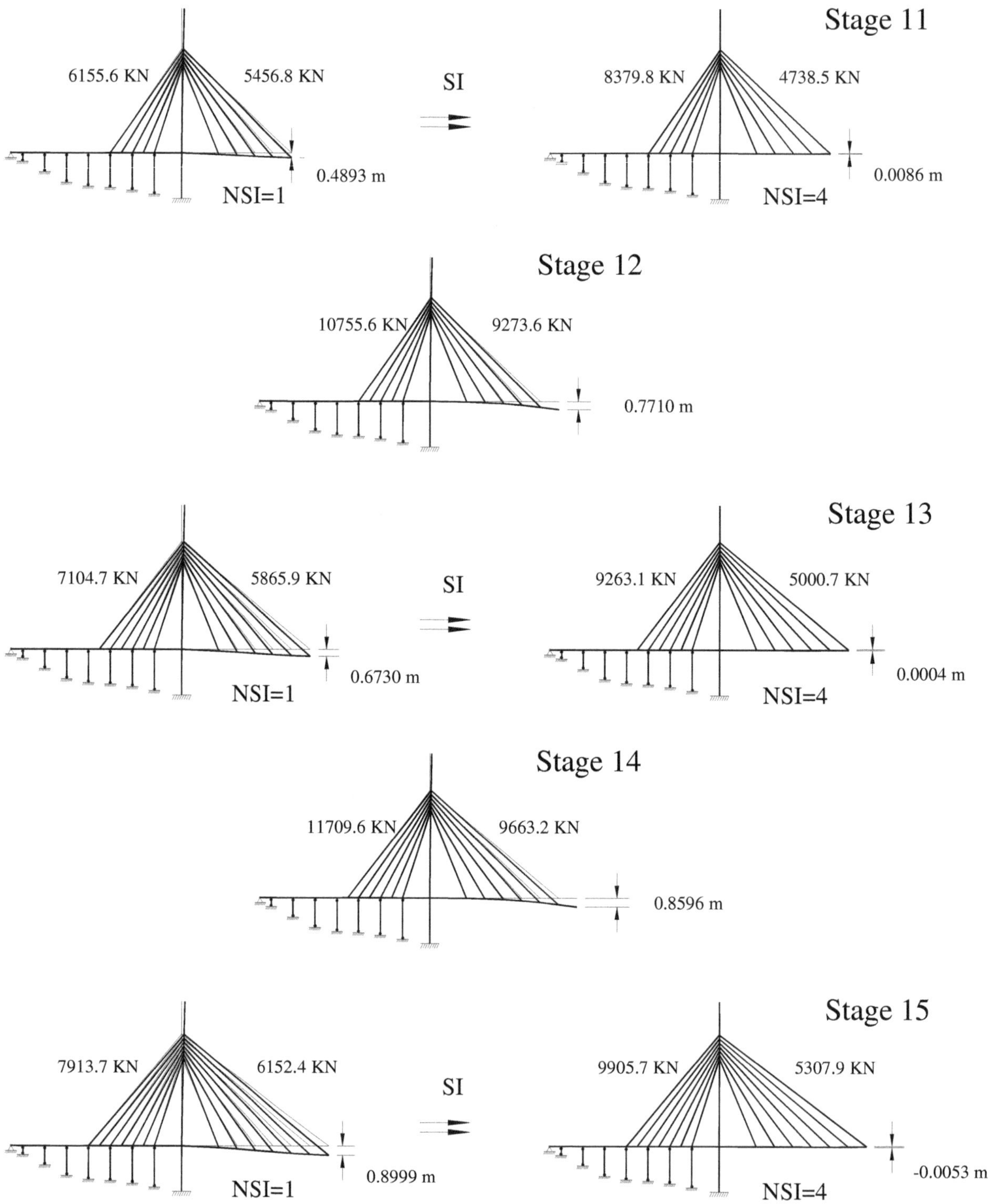
Stage 11
6155.6 KN
5456.8 KN
0.4893 m
NSI=1
SI
8379.8 KN
4738.5 KN
0.0086 m
NSI=4
Stage 12
10755.6 KN
9273.6 KN
0.7710 m
Stage 13
7104.7 KN
5865.9 KN
0.6730 m
NSI=1
SI
9263.1 KN
5000.7 KN
0.0004 m
NSI=4
Stage 14
11709.6 KN
9663.2 KN
0.8596 m
Stage 15
7913.7 KN
6152.4 KN
0.8999 m
NSI=1
SI
9905.7 KN
5307.9 KN
-0.0053 m
NSI=4

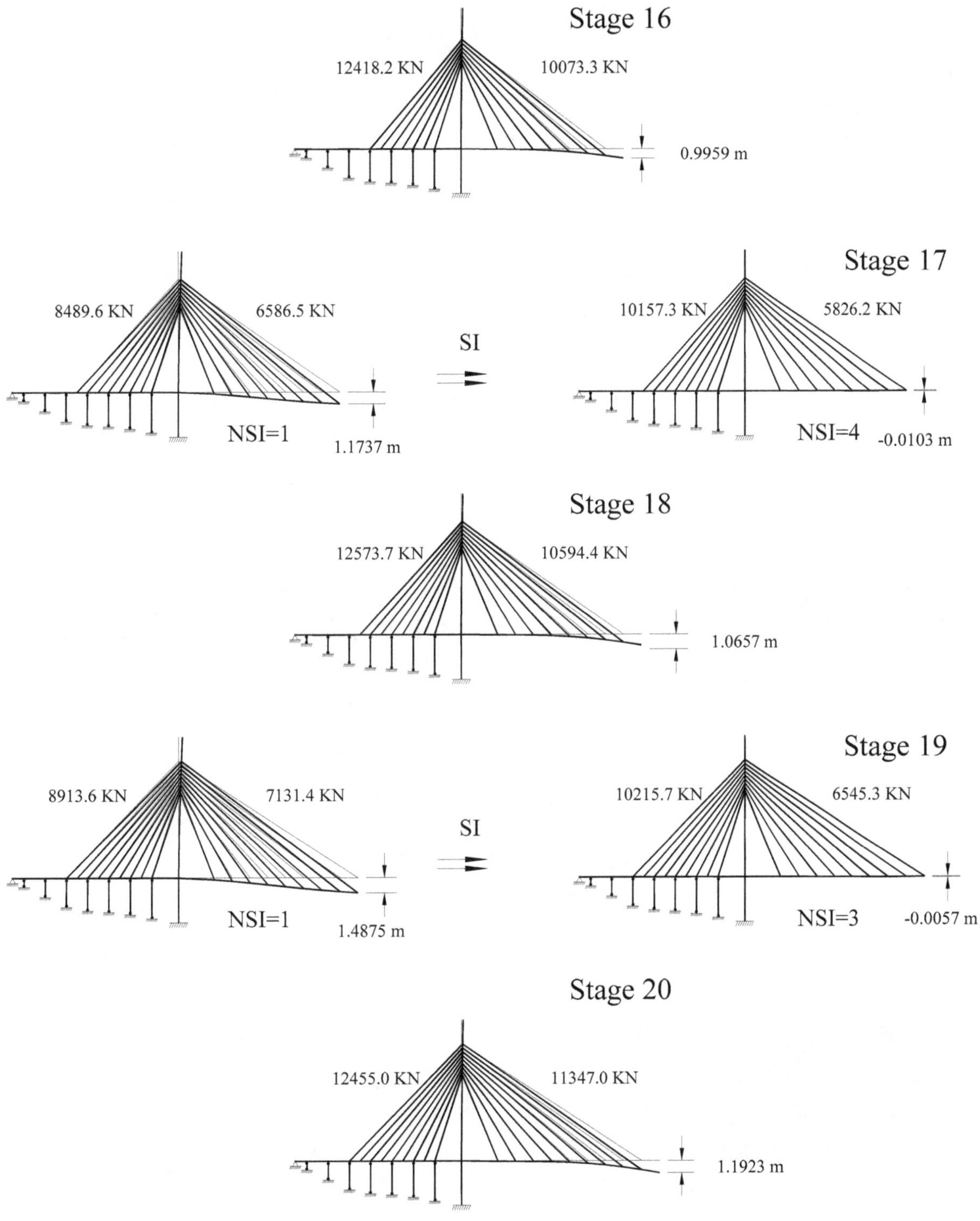
Stage 16
12418.2 KN
10073.3 KN
0.9959 m
Stage 17
8489.6 KN
6586.5 KN
SI
NSI=1
1.1737 m
10157.3 KN
5826.2 KN
NSI=4
-0.0103 m
Stage 18
12573.7 KN
10594.4 KN
1.0657 m
Stage 19
8913.6 KN
7131.4 KN
SI
NSI=1
1.4875 m
10215.7 KN
6545.3 KN
NSI=3
-0.0057 m
Stage 20
12455.0 KN
11347.0 KN
1.1923 m

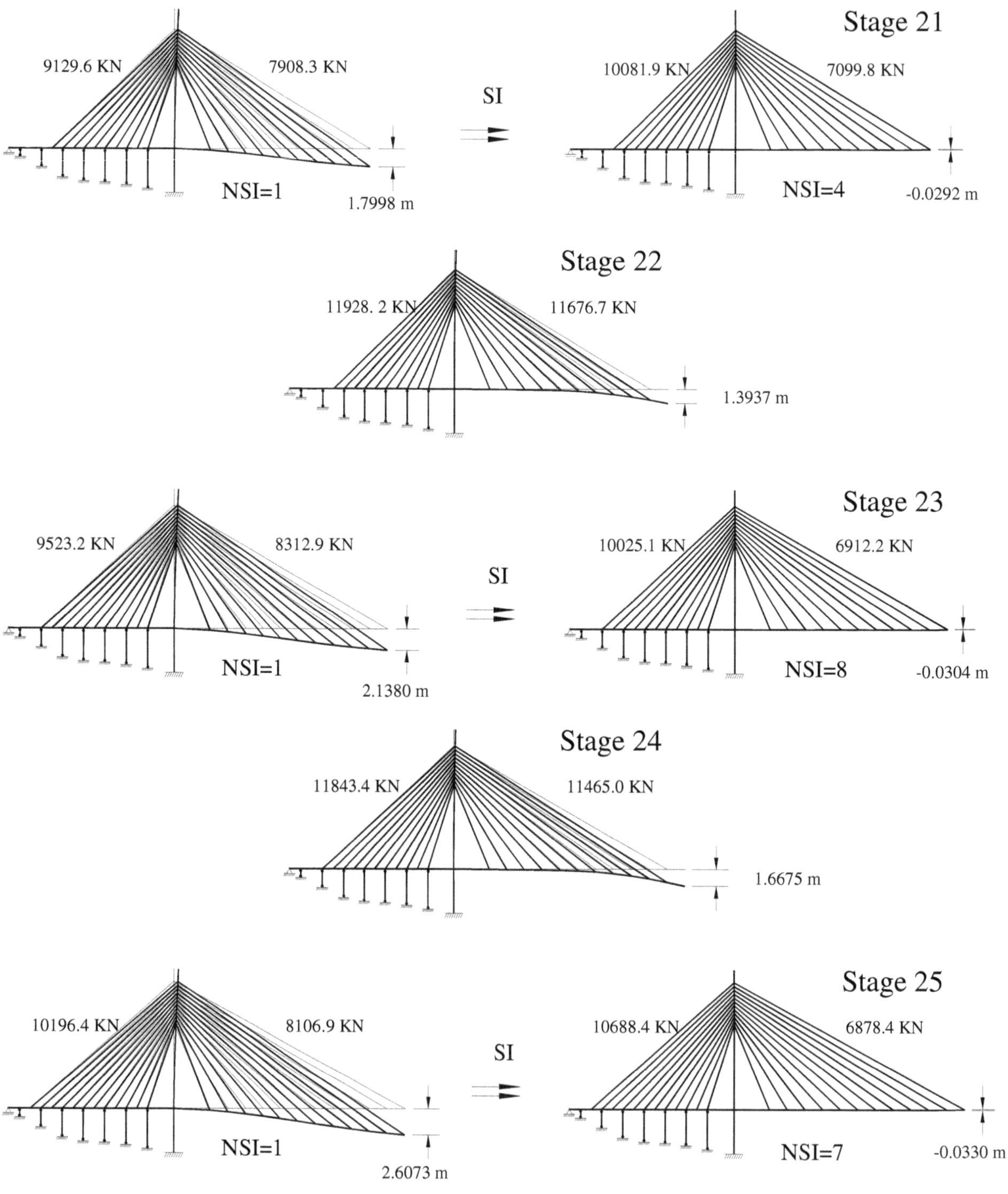
Stage 21
9129.6 KN
7908.3 KN
NSI=1
1.7998 m
SI
10081.9 KN
7099.8 KN
NSI=4
-0.0292 m
Stage 22
11928. 2 KN
11676.7 KN
1.3937 m
Stage 23
9523.2 KN
8312.9 KN
NSI=1
2.1380 m
SI
10025.1 KN
6912.2 KN
NSI=8
-0.0304 m
Stage 24
11843.4 KN
11465.0 KN
1.6675 m
Stage 25
10196.4 KN
8106.9 KN
NSI=1
2.6073 m
SI
10688.4 KN
6878.4 KN
NSI=7
-0.0330 m

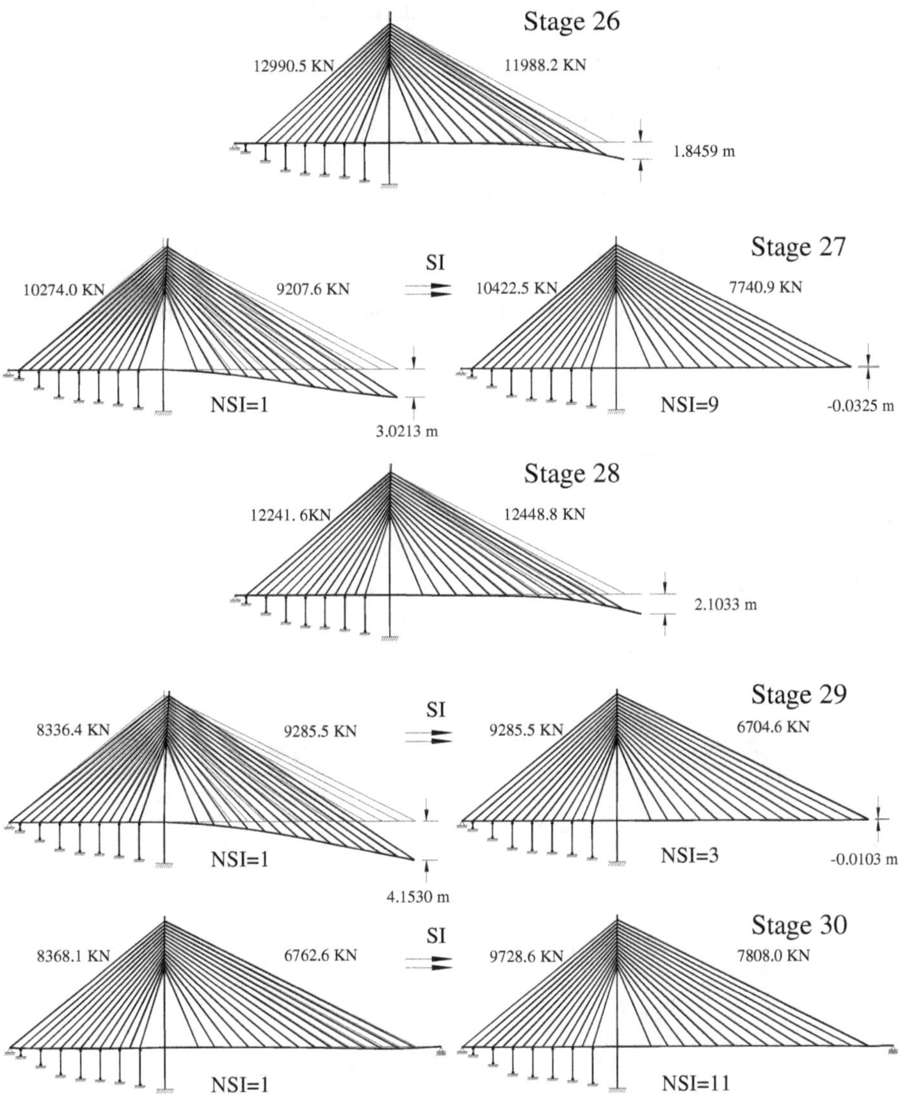

Figure 9. Initial shape of the Kao Ping Hsi Bridge at each erection stage based on forward process analysis.

exterior stay cables are installed and stressed. Under these conditions, the new equilibrium position can be estimated by the shape iteration. The girders can therefore be kept in a horizontal position and the corresponding bending moments can also be reduced. At stage 30, the shape iteration has to be conducted to ensure that the designed shape of the whole bridge for the construction completion is almost identical to that from the whole bridge analysis. This is because the initial shapes of the bridge at erection stages by the shape iteration associated with the estimated initial cable forces are not unique. In summary, Figure 9 provides the necessary data for checking and controlling the real-time erection procedure of the bridge during construction.

4.3. Backward process analysis

According to the backward process analysis of cable-stayed bridges during construction by the single cantilever method in Section 3.3, the initial shape of the Kao Ping Hsi Bridge at each erection stage for a total of 30 stages is obtained, as shown in Figure 10. The geometric configuration of the bridge and the corresponding vertical displacement of the girder tip in the main span as well as the element forces of the exterior stay cables are illustrated at each erection stage. No shape iteration is needed for the backward process analysis, implying that the initial shapes of the bridge at erection stages are unique. The computation is started

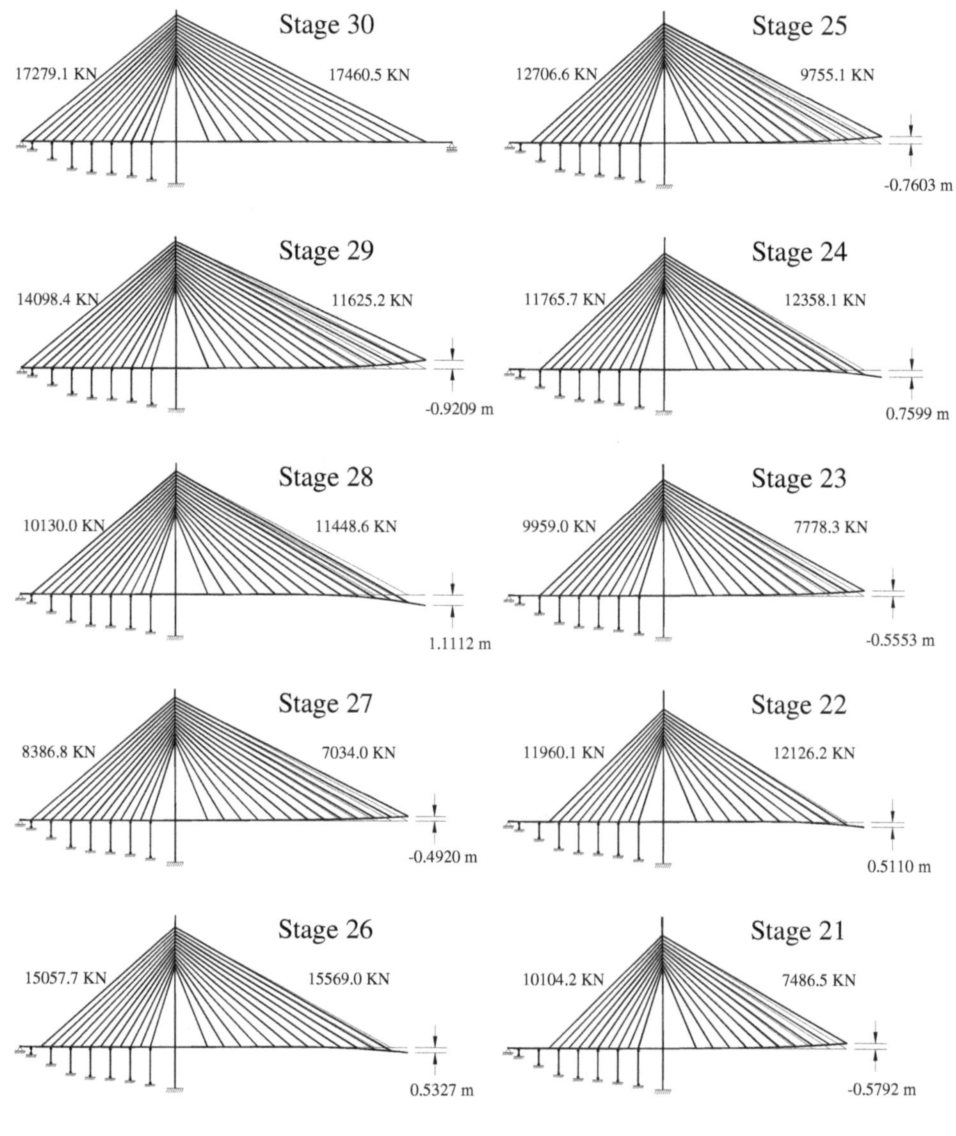

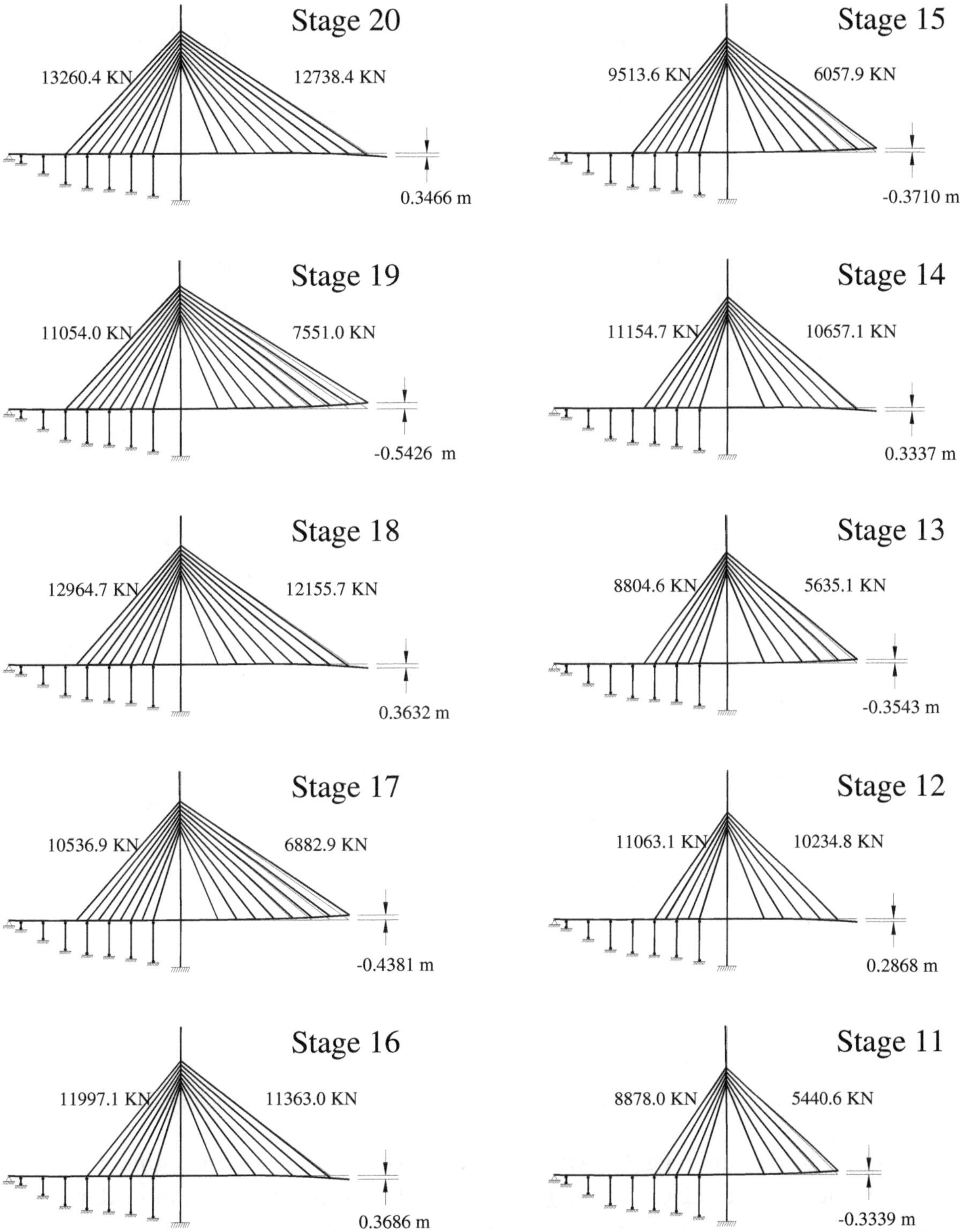
Stage 20
13260.4 KN
12738.4 KN
0.3466 m
Stage 15
9513.6 KN
6057.9 KN
-0.3710 m
Stage 19
11054.0 KN
7551.0 KN
-0.5426 m
Stage 14
11154.7 KN
10657.1 KN
0.3337 m
Stage 18
12964.7 KN
12155.7 KN
0.3632 m
Stage 13
8804.6 KN
5635.1 KN
-0.3543 m
Stage 17
10536.9 KN
6882.9 KN
-0.4381 m
Stage 12
11063.1 KN
10234.8 KN
0.2868 m
Stage 16
11997.1 KN
11363.0 KN
0.3686 m
Stage 11
8878.0 KN
5440.6 KN
-0.3339 m

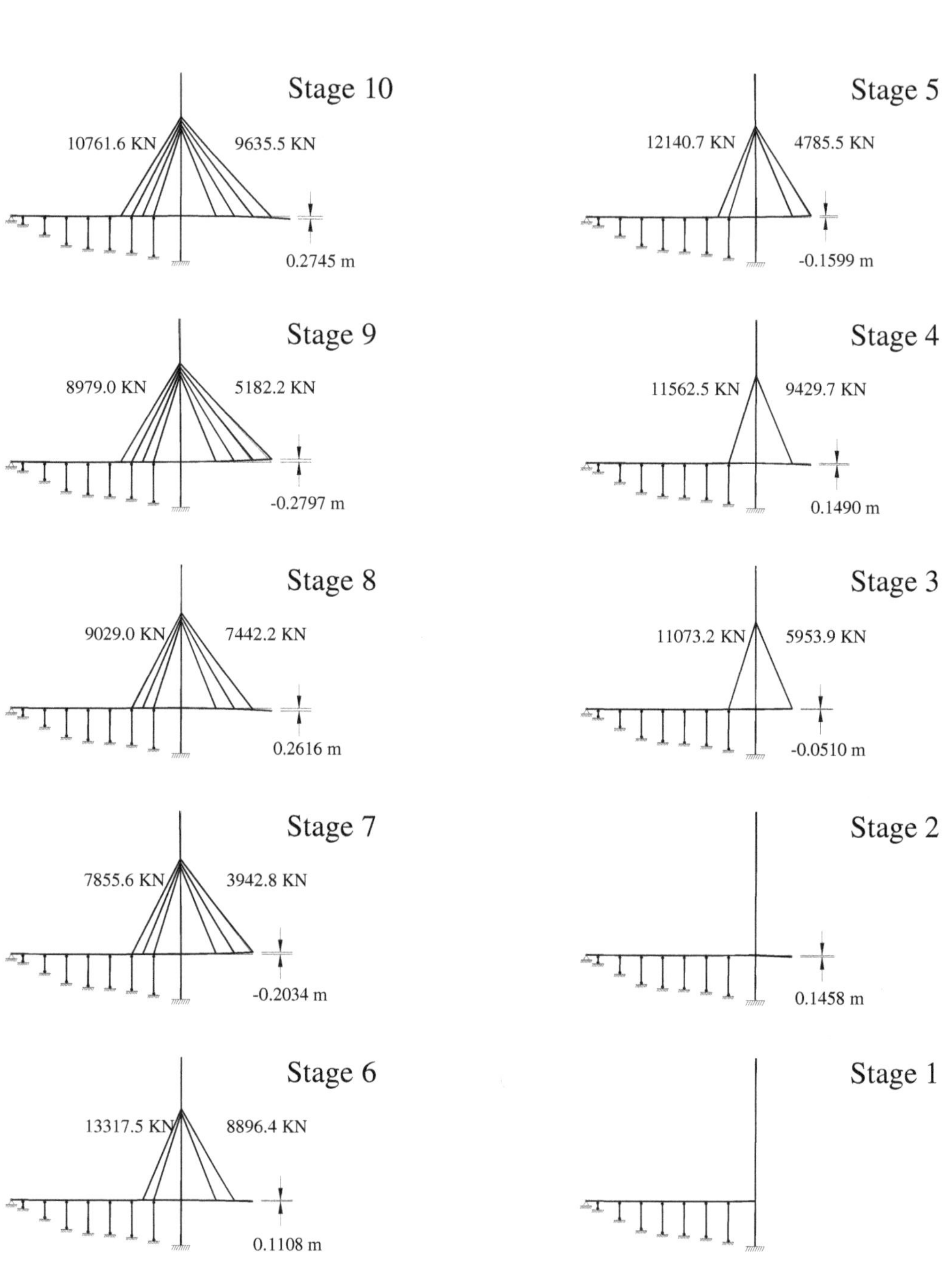

Figure 10. Initial shape of the Kao Ping Hsi Bridge at each erection stage based on backward process analysis.

with the whole bridge at stage 30. After removing the existing girder segments at erection stages with odd number (29, 27, 25,...), the upward displacement of the girder tip in the main span occur, which can be considered as the precamber of the girder for the next erection stage. Such precamber can keep the girders in the original designed configuration as well as reduce the bending moments of the girders. After removing the existing exterior stay cables at erection stages with even number (28, 26, 24,...), the large downward displacement of the girder tip in the main span occur and the corresponding bending moments become quite large. In summary, Figure 10 provides the necessary data for checking and controlling the real-time erection procedure of the bridge during construction.

5. Conclusions

The objective of this chapter is to fully understand the system behaviors with the appropriate initial shapes of cable-stayed bridges at each erection stage during construction by the cantilever method. Two computational procedures during erection stages: a forward process analysis and a backward process analysis, are presented for this reason. On the basis of the two procedures, a series of initial shape analyses are conducted to investigate the bridge shapes at each erection stage. Numerical examples based on finite element models of the Kao Ping Hsi Bridge in Taiwan are developed to validate the two proposed approaches. Based on the numerical analysis in this study, some conclusions are made as follows:

1. Both the forward process analysis and the backward process analysis provide the necessary data for checking and controlling the real-time erection procedure of a cable-stayed bridge during construction. The designed shape, i.e., the geometric configuration and the prestress distribution, of the whole bridge can then be achieved.
2. The advantage of the forward process analysis is that the real-time factors of a cable-stayed bridge, such as creep and shrinkage of concrete, any alteration in design, etc., can be taken into consideration during construction. However, the shape iteration at the final erection stage has to be conducted to ensure that the designed shape of the whole bridge for the construction completion is almost identical to that from the whole bridge analysis. This is because the initial shapes of the bridge at erection stages by the shape iteration associated with the estimated initial cable forces are not unique.
3. The advantage of the backward process analysis is that the initial shape of a cable-stayed bridge at each erection stage can be obtained by the equilibrium iteration without the shape iteration. Such initial shape determined by the equilibrium conditions can be considered as the desired correct position of the bridge for the next erection stage, in which the girder is precambered upwards. Under these conditions, the initial shapes of the bridge at erection stages are unique. In other words, the computational efficiency of the backward process analysis without the shape iteration is better than that of the forward process analysis with the shape iteration. In contrast, the disadvantage of the backward process analysis is that the real-time factors of the bridge, such as creep and shrinkage of concrete, any alteration in design, etc., can not be taken into consideration during construction due to the fact that the computation is performed backwards from the whole bridge.
4. Both the forward process analysis and the backward process analysis of cable-stayed bridges during construction by the single cantilever method based on the sequence of erection stages are presented in this study. These concepts can also be applied to the double cantilever method.

Acknowledgements

The authors would like to acknowledge the support of the National Science Council, Taiwan, under Grant No. NSC 89-2211-E-033-009. Parts of the chapter are produced from the authors' previous publication [11].

Author details

Ming-Yi Liu* and Pao-Hsii Wang

*Address all correspondence to: myliu@cycu.edu.tw

Department of Civil Engineering, Chung Yuan Christian University, Jhongli City, Taoyuan County, Taiwan

References

[1] Troitsky, M.S. (1988). "Cable-stayed bridges: Theory and design." Second Edition, Blackwell Scientific Publications, Oxford, UK.

[2] Walther, R., Houriet, B., Isler, W., Moïa, P., and Klein, J.F. (1999). "Cable stayed bridges." Second Edition, Thomas Telford, Ltd, London, UK.

[3] Gimsing, N.J., and Georgakis, C.T. (2012). "Cable supported bridges: Concept and design." Third Edition, John Wiley & Sons, Ltd, Chichester, UK.

[4] Leonhardt, F., and Zellner, W. (1991). "Past, present and future of cable-stayed bridges." *Proceedings of the Seminar - Cable-Stayed Bridges: Recent Developments and their Future*, Yokohama, Japan, 1-33, December 10-11.

[5] Virlogeux, M. (1991). "Erection of cable-stayed bridges." *Proceedings of the Seminar - Cable-Stayed Bridges: Recent Developments and their Future*, Yokohama, Japan, 77-106, December 10-11.

[6] Nakagawa, T., Sugii, K., Nagai, M., and Mochizuki, H. (2000). "Design and erection of a cable trussed bridge." *Proceedings of the Thirteenth KKNN Symposium on Civil Engineering*, Taipei, Taiwan, 107-112, December 7-8.

[7] Wang, P.H., Tseng, T.C., and Yang, C.G. (1993). "Initial shape of cable-stayed bridges." *Computers and Structures*, 46(6), 1095-1106.

[8] Wang, P.H., and Yang, C.G. (1996). "Parametric studies on cable-stayed bridges." *Computers and Structures*, 60(2), 243-260.

[9] Wang, P.H., Lin, H.T., and Tang, T.Y. (2002). "Study on nonlinear analysis of a highly redundant cable-stayed bridge." *Computers and Structures*, 80(2), 165-182.

[10] Wang, P.H., Tang, T.Y., and Zheng, H.N. (2004). "Analysis of cable-stayed bridges during construction by cantilever methods." *Computers and Structures*, 82(4-5), 329-346.

[11] Liu, M.Y., and Wang, P.H. (2012). "Finite element analysis of cable-stayed bridges with appropriate initial shapes under seismic excitations focusing on deck-stay interaction." In Sezen, H. (Ed.), Chapter 9 of *Earthquake Engineering* (pp. 231-256). InTech, Rijeka, Croatia. ISBN: 978-953-51-0694-4.

[12] Cheng, W.L. (2001). "Kao Ping Hsi Bridge." Taiwan Area National Expressway Engineering Bureau, Ministry of Transportation and Communications, Taipei, Taiwan.

[13] Ernst, H.J. (1965). "Der E-Modul von Seilen unter Berücksichtigung des Durchhanges." *Der Bauingenieur*, 40(2), 52-55. (in German).

[14] Fleming, J.F. (1979). "Nonlinear static analysis of cable-stayed bridge structures." *Computers and Structures*, 10(4), 621-635.

5

Detection of Accelerating Transient of Aseismic Rock Strain using Precursory Decline in Groundwater Radon

Ming-Ching T. Kuo

1. Introduction

A seismic slip can be preceded by accelerating aseismic slips near the hypocenter of an impending earthquake. Compared with the strain step recorded at the time of the earthquake, the precursory strain from aseismic fault slips are transient and small. It is of practical interest to be able to detect the accelerating aseismic rock strain for the warning of local disastrous earthquakes. From a series of laboratory experiments in un-drained conditions, I discovered a significant drop in groundwater radon, greater than 50 %, by a mechanism of radon volatilization into the vapor phase. Both radon volatilization and rock dilatancy offer attractive in-situ physical mechanisms for a premonitory decrease in groundwater radon. In addition, we have been monitoring groundwater radon at well (D1) in the Antung hot spring in eastern Taiwan since July 2003 (Fig. 1). The Antung hot spring is located near the Chihshang fault that ruptured during two 1951 earthquakes of magnitudes M=6.2 and M=7.0 (Hsu, 1972). Based on the long-term observation at the Antung hot spring, I discovered that an un-drained brittle aquifer near an active fault can be used as a natural strain meter for detecting recurrent precursory radon declines. The observed radon minimum decreases as the earthquake magnitude increases. The un-drained condition at well (D1) is essential for the development of two phases (vapor and water) in the rock cracks, which is attributed to the happening of recurrent precursory radon declines. The concurrent concentration declines in groundwater-dissolved gases (radon, methane, and ethane) support the un-drained condition at well (D1) and the mechanism of in-situ volatilization of groundwater-dissolved gases. Radon precursory declines in groundwater can only be detected in certain locations with favorable geological conditions. An un-drained brittle aquifer near an active fault is recommended to monitor recurrent precursory radon declines. The objective of this chapter is to provide a practical means to detect the accelerating aseismic rock strain using the precursory decline in groundwater radon. A case study is provided in this chapter to illustrate the application of an undrained brittle aquifer as a natural strain meter to detect the accelerating transient of aseismic rock strain by monitoring precursory decline in groundwater radon.

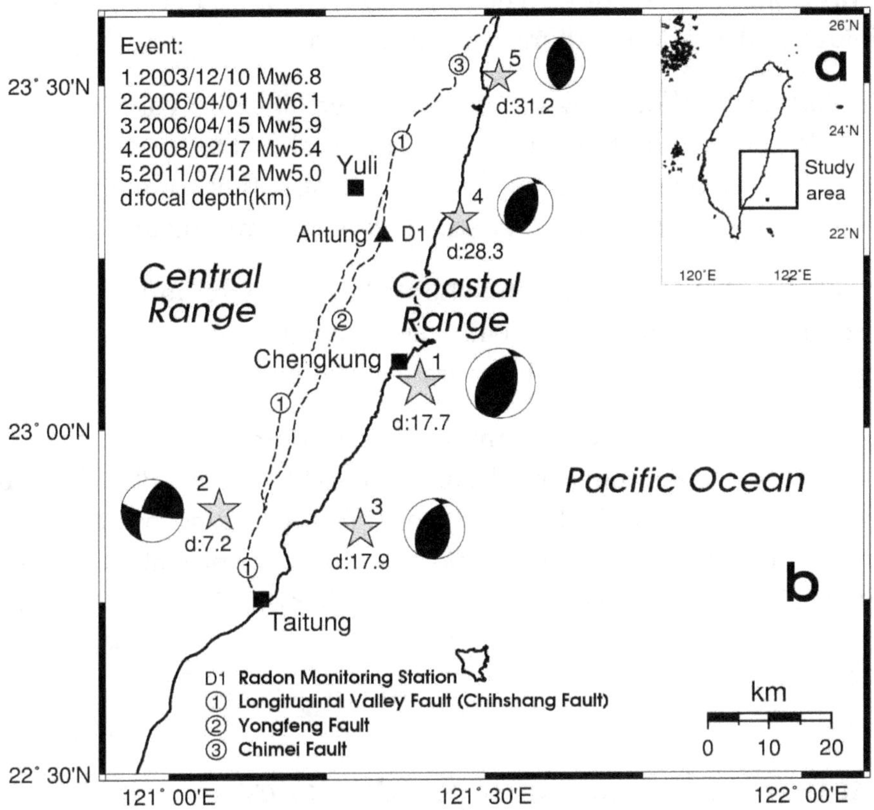

Figure 1. Map of the epicenters of the earthquakes that occurred on December 10, 2003, April 1 and 15, 2006, February 17, 2008, July 12, 2011 near the Antung hot spring. (a) Geographical location of Taiwan. (b) Study area near the Antung hot spring (filled stars: mainshocks, filled triangle: radon-monitoring station).

2. Monitoring methods

We have been monitoring groundwater radon at a well (D1) located at the Antung hot spring in eastern Taiwan since July 2003. Figure 1 shows that the radon-monitoring well (D1) is about 3 km southeast of the Chihshang fault. Discrete samples of groundwater were pumped and collected from the radon-monitoring well (D1) twice per week for analysis of radon content. We also started to analyze methane and ethane from November 2007 and November 2010, respectively.

The production interval of the radon-monitoring well (D1) ranges from 167 m to 187 m. Water samples were collected in a 40 mL glass vial with a TEFLON-lined cap. To obtain representative water samples for radon, methane, and ethane measurements, a minimum purge of three well-bored volumes were required before taking samples. A minimum of 50 min purging-time was required with a pumping rate at around 200 L/min.

It is important to ensure the radon not to escape during the sampling procedure and the sample transportation and preparation. For every sampling, the sample vial was inverted to check if any bubbles were present in the vial. The sample water with gas bubbles present was discarded and sampling was repeated. The sampling time of sample collection were recorded and radon measurement was done within 4 days. The samples were stored and transported in a cooler to minimize biological degradation of methane and ethane.

The liquid scintillation method was applied to determine the concentration of radon in groundwater [12]. Radon was first extracted into a mineral-oil based scintillation cocktail from the water samples, and then measured with a liquid scintillation counter (LSC). The measurement results were corrected for the radon decay between sampling and counting.

Calibration factor for the LSC measurements should be at least 6 cpm/pCi with the background not exceeding 6 cpm. For a count time of 50 min and background less than 6 cpm, we achieved a detection limit below 18 pCi/L using the sample volume of 15 ml.

The head space method was used to determine the concentrations of methane and ethane in groundwater with a gas chromatograph (Shimadzu GC-14A), a HP-Plot/Q, 30 m, 0.53 mm i.d. capillary column, and a flame ionization detector (FID). The concentrations of methane and ethane measured in the head space were converted to groundwater concentrations using Henry's constants of 31.5 and 23.9, respectively, for methane and ethane at a room temperature of 27 ℃.

3. In-situ radon-volatilization and rock-dilatancy

Radon partitioning into the gas phase can explain the anomalous decreases of radon concentration in groundwater precursory to the earthquakes [7]. Based on radon phase-behavior and rock-dilatancy process [1], [8] developed a mechanistic model to correlate the observed decline in radon with the rock strain precursory to an earthquake. We will present the model with two parts, i.e., the radon-volatilization model and the rock-dilatancy model. The radon-volatilization model which correlates the radon decline to the gas saturation can be expressed as follows.

$$C_0 = C_w (H S_g + 1) \tag{1}$$

where C_0 is the stabilized radon concentration in groundwater before a radon anomaly, pCi/L; C_w is the observed radon decline in groundwater during a radon anomaly, pCi/L; S_g is the gas saturation in aquifer during a radon anomaly, fraction; H is Henry's constant for radon, dimensionless. The Henry's constant (H) at formation temperature (60 ℃) is 7.91 for radon [5].

The rock-dilatancy model which correlates the rock strain to the gas saturation can be expressed as follows.

$$d\varepsilon \cong \phi \; S_g \tag{2}$$

where $d\varepsilon$ is the rock strain precursory to an earthquake, fraction; ϕ is the fracture porosity of aquifer, fraction; S_g is the gas saturation in aquifer during an radon anomaly, fraction.

Based on the radon volatilization and rock dilatancy models, we can correlate the radon decline in groundwater radon to the rock strain precursory to an earthquake. Combining equations (1) and (2), we obtain equation (3) as follows.

$$d\varepsilon \cong \frac{\phi}{H}\left(\frac{C_0}{C_w}-1\right) \tag{3}$$

where $\left(\frac{C_0}{C_w}-1\right)$ is normalized radon decline precursory to an earthquake, dimensionless. Given the precursory decline in groundwater radon such as, Figure 2, equation (3) can be used to calculate the precursory crustal-strain transient from aseismic fault slips. Case studies are provided to illustrate the application of an un-drained brittle aquifer as a natural strain meter to detect the accelerating transient of aseismic rock strain by observing premonitory decline in groundwater radon.

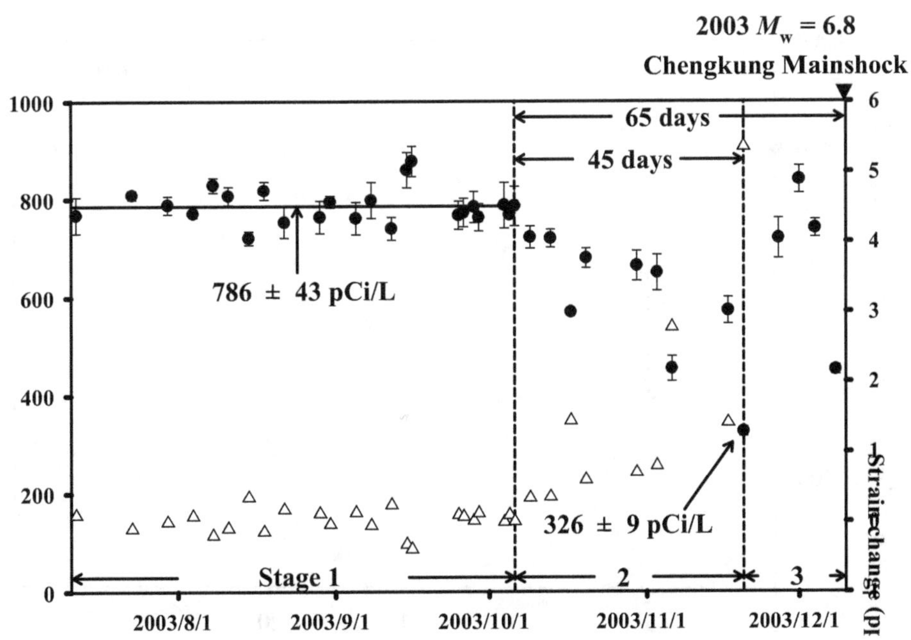

Figure 2. Observed radon decline and calculated crustal-strain transient prior to 2003 M_W 6.8 Chengkung earthquake at the monitoring well (D1) in the Antung hot spring (solid circles: observed radon concentration; open triangles: calculated crustal-strain). Stage 1 is buildup of elastic strain. Stage 2 is development of cracks and gas saturation. Stage 3 is influx of groundwater.

4. A case study

[7] discovered that a significant drop in groundwater radon, greater than 50 %, can be generated in the laboratory by a process of radon volatilization into the gas phase in un-drained conditions. We conducted a series of radon-partitioning experiments to determine the variation of the radon concentration remaining in groundwater at various levels of gas saturation. Figure 3 shows the results of vapor-liquid, two-phase radon-partitioning experiments conducted at formation temperature (60 ℃) using formation brine from the Antung hot spring. The processes of rock dilatancy, under-saturation and radon volatilization offer an attractive mechanism to monitor anomalous radon declines in groundwater radon precursory to an earthquake. Given an observed radon decline precursory to an earthquake, we can apply Figure 3 to estimate the amount of gas saturation in micro-cracks developed in aquifer during an radon anomaly. For example, a gas saturation of 10 % in cracks developed in the aquifer rock when the radon concentration in groundwater decreased from 780 pCi/L to 330 pCi/L precursory to the 2003 M_w 6.8 Chengkung earthquake. [16] reported that the fracture porosity for naturally fractured rocks ranges from 0.00008 to 0.0003. To generate an in-situ gas saturation of 10% in a fractured aquifer, a crustal strain of 8.0 ppm and 30.0 ppm is required for a fracture porosity of 0.00008 and 0.0003, respectively. Both low-porosity and un-drained conditions are favorable for applying a fractured aquifer as a natural strain meter by monitoring precursory decline in groundwater radon. It is of practical interest to be able to detect the accelerating aseismic rock strains by monitoring the radon concentration in groundwater.

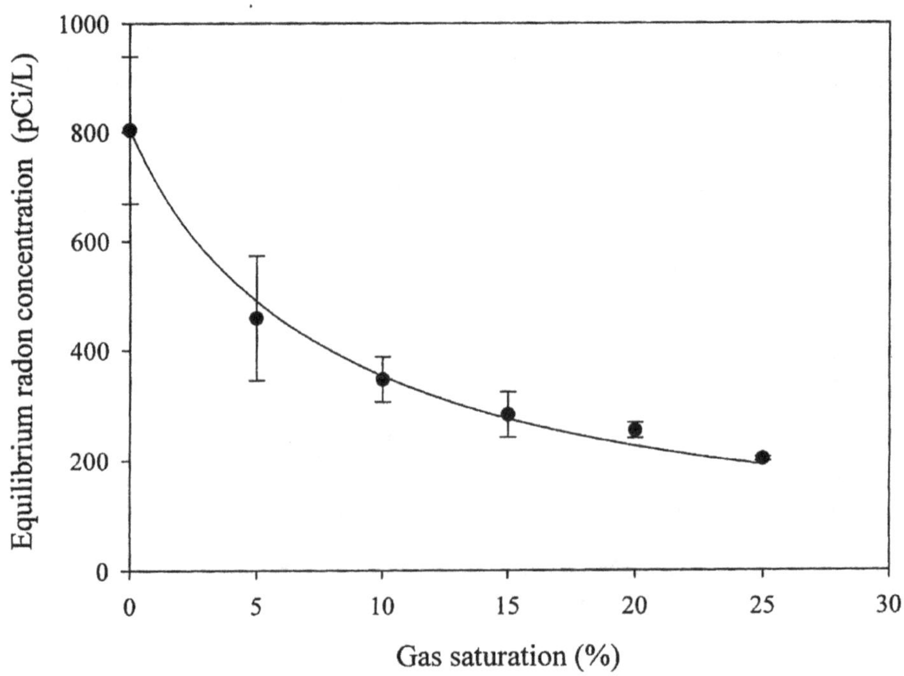

Figure 3. Variation of radon concentration remaining in groundwater with gas saturation at 60℃ using formation brine from the Antung hot spring.

Since July 2003, we have observed recurrent recurrent anomalous declines in groundwater radon at the Antung well (D1) in eastern Taiwan precursory to the 2003 M_w = 6.8 Chengkung, 2006 M_w = 6.1 Taitung, 2008 M_w = 5.4 Antung, and 2011 M_w = 5.0 Chimei earthquakes that occurred on December 10, 2003, April 1, 2006, February 17, 2008, and July 12, 2011, respectively. The epicenters of the 2003 M_w = 6.8, 2006 M_w = 6.1, 2008 M_w = 5.4, and 2011 M_w = 5.0 earthquakes were located only 24 km, 52 km, 13 km, and 32 km, respectively, from the observation well (D1).

The observed radon anomalies can be correlated with the magnitude and precursor time of upcoming earthquakes. We define the precursor time for radon as the time interval between the moment when the trend of the radon concentration starts to decline and the time of occurrence of the earthquake. Based on the radon anomalies observed prior to (1) 2003 M_w 6.8 Chengkung, (2) 2006 M_w 6.1 Taitung, (3) 2008 M_w 5.4 Antung, and (4) 2011 M_w 5.0 Chimei earthquakes, Table 1 summarizes the precursor time and radon minima. [11] shows that as the magnitude of earthquakes increases, the precursor time for radon anomalies increases and the observed radon minima decrease. Monitoring precursory decline in groundwater radon at a suitable geological site can be a useful means of forecasting the magnitude and precursor time of local disastrous earthquakes.

Earthquake	Moment magnitude, M_W, *(dimensionless)*	Precursory time (day)	Radon minimum (pCi/L)
2003 Chengkung	6.8	65	326 ± 9
2006 Taitung	6.1	61	371 ± 9
2008 Antung	5.4	56	480 ± 43
2011 Chimei	5.0	54	447 ± 18

Table 1. Observed precursory time and radon minimum at well (D1) prior to (1) 2003 M_W = 6.8 Chengkung, (2) 2006 M_W = 6.1 Taitung, (3) 2008 M_W = 5.4 Antung, and (4) 2011 M_W = 5.0 Chimei earthquakes.

[4] show that the Antung hot spring situated in an andesitic block and surrounded by a ductile mudstone of the Lichi mélange. Figure 4 shows the geological map and cross section near well (D1) in the area of Antung hot spring which is a low-porosity fractured confined aquifer. The groundwater is in un-drained conditions at well (D1). [13] and [15] suggested that the development of new cracks in aquifer rock could occur at a rate faster than the recharge of pore water in un-drained conditions. Gas saturation developed in the rock cracks and groundwater-dissolved radon then volatilized into the gas phase. We also observed simultaneous anomalous declines in groundwater-dissolved radon and methane precursory to the 2008 M_w = 5.4 Antung earthquake [9]. The mechanism of in-situ radon volatilization was substantiated.

The composition of groundwater-dissolved gases taken from a separator flow test at well (D1) on December 26, 2006 consists of 62.8 % of nitrogen, 36.7 % of methane, and 0.5 % of ethane

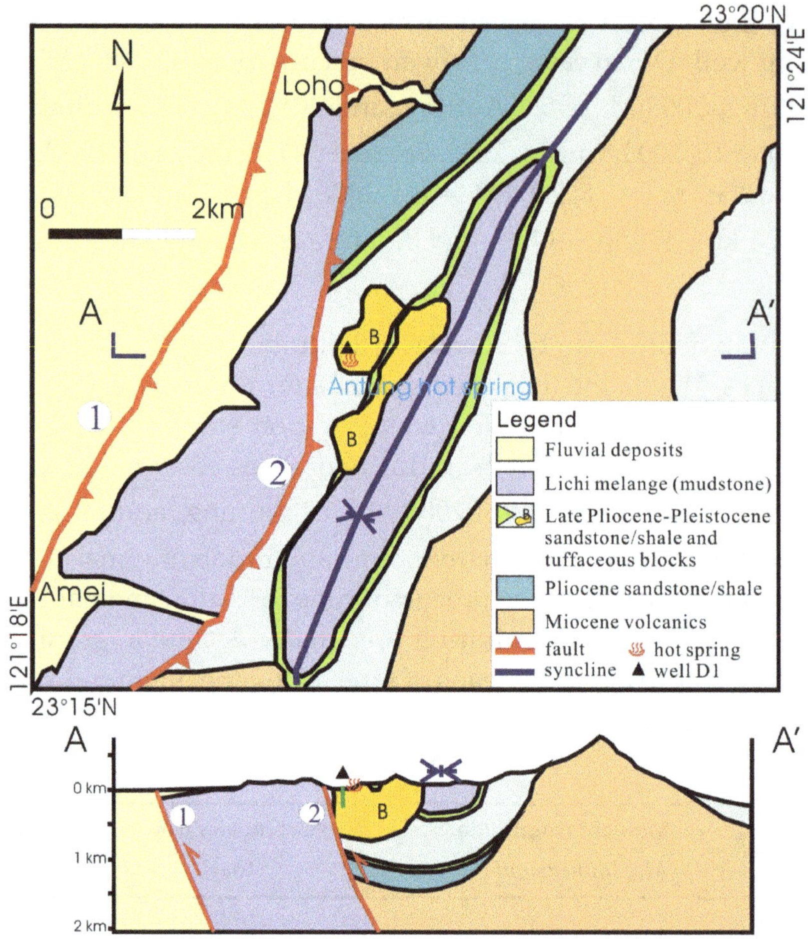

Figure 4. Geological map and cross section near the radon-monitoring well (D1) in the area of Antung hot spring. (B: tuffaceous andesitic blocks; 1): Chihshang, or, Longitudinal Valley Fault, 2): Yongfeng Fault)

by volume. In addition to radon and methane, we initiated the monitoring of groundwater-dissolved ethane at well (D1) in the Antung hot spring since November 30, 2010 to corroborate the in-situ volatilization mechanism. The in-situ radon-volatilization model for groundwater-dissolved radon, methane, and ethane can be expressed as follows.

$$C_{0,Rn} = C_{w,Rn}(H_{Rn}S_g + 1) \tag{4}$$

$$C_{0,Me} = C_{w,Me}(H_{Me}S_g + 1) \tag{5}$$

$$C_{0,Et} = C_{w,Et}(H_{Et}S_g + 1) \tag{6}$$

where $C_{0,Rn}$ is the stabilized radon concentration in groundwater before a radon anomaly, pCi/L; $C_{w,Rn}$ is the observed radon decline in groundwater during a radon anomaly, pCi/L; S_g is the gas saturation in aquifer during a radon anomaly, fraction; H_{Rn} is Henry's constant for radon, dimensionless; $C_{0,Me}$ is the stabilized methane concentration in groundwater before a methane anomaly, mg/L; $C_{w,Me}$ is the observed methane decline in groundwater during a methane anomaly, mg/L; H_{Me} is Henry's constant for methane, dimensionless; $C_{0,Et}$ is the stabilized ethane concentration in groundwater before a ethane anomaly, mg/L; $C_{w,Et}$ is the observed ethane decline in groundwater during a ethane anomaly, mg/L; H_{Et} is Henry's constant for ethane, dimensionless. The Henry's coefficients at 60 °C are 7.91, 37.6, and 38.2 for radon, methane, and ethane, respectively. According to equations (4), (5), and (6), the mechanism of in-situ volatilization predicts the concurrent concentration declines in groundwater-dissolved radon, methane, and ethane.

Prior to the 2011 Chimei earthquake, the concurrent concentration declines in groundwater-dissolved radon, methane, and ethane were observed [10]. Figure 5 shows the observed concentration anomalies for radon, methane, and ethane, respectively, precursory to the 2011 Chimei earthquake. The concentration errors are ±1 standard deviation after simple averaging of triplicates. Radon, methane, and ethane decreased from background levels of 752 ± 24 pCi/L, 8.24 ± 0.48 mg/L, and 0.217 ± 0.010 mg/L to minima of 447 ± 18 pCi/L, 5.81 ± 0.30 mg/L, and 0.161 ± 0.008 mg/L, respectively (Figure 5). The mechanism of in-situ radon volatilization was confirmed again by the simultaneous anomalous declines in groundwater-dissolved radon, methane, and ethane.

The anomalous decline of radon concentration in groundwater was observed prior to the 2003 Chengkung earthquake. Figure 2 shows that the sequence of events can be divided into three stages. During Stage 1 (from July 2003 to September 2003), radon concentration in groundwater was fairly stable (around 780 pCi/L). During Stage 1, there was a slow, steady increase of effective stress and an accumulation of tectonic strain. Sixty-five days before the 2003 $M_w = 6.8$ Chengkung earthquake which occurred on December 10, 2003, the concentration of radon started to decrease and reached a minimum value of 330 pCi/L twenty days before the earthquake. We define this 45-day period as Stage 2. Dilation of the rock mass occurred during Stage 2. When the aquifer is in un-drained conditions, the development of new cracks in aquifer rock could occur at a rate faster than the recharge of pore water. Gas saturation then developed in the rock cracks and groundwater-dissolved radon volatilized into the gas phase. After the minimum point of radon concentration (Stage 3), groundwater continued to encroach into the rock cracks and the water saturation in the aquifer began to increase. During Stage 3, groundwater-dissolved radon increased and recovered to the background level. The main shock produced a sharp coseismic anomalous decrease (~300 pCi/L).

The 2003 Chengkung earthquake's dislocation fault model was analyzed by Wu et al using a computer code by [14]. [17] determined the fault geometry utilizing aftershock distribution and geology (CGS 2000a, b) and assumed a thrust fault parallel to Coastal Range with strike N20°E, with a bend at a depth of 18 km. The fault-plane dips 60°SE and 45°SE above and below 18 km respectively. Assume both the lower and upper fault-planes extend a maximum of 33 km from north to south. Rupturing of the lower and upper fault-planes occurred within depths of 18-36 km and 5-18 km, respectively. For an optimal fit with the coseismic ground deformation, the lower fault slipped 61.6 cm with a rake of 81.7° and the upper fault slipped 26 cm with

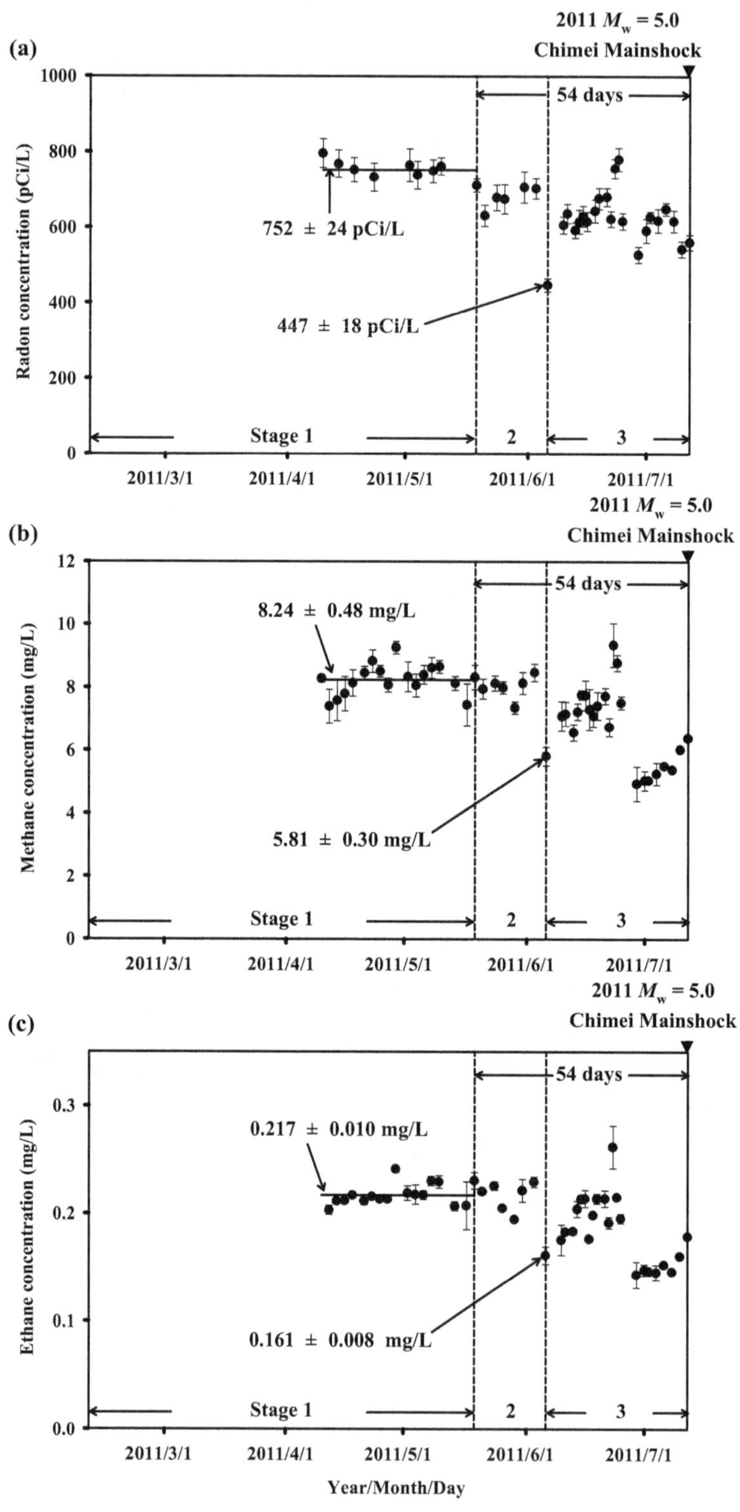

Figure 5. Observed concentration anomalies (a) radon, (b) methane, and (c) ethane prior to 2011 M_W 5.0 Chimei earthquake. Stage 1 is buildup of elastic strain. Stage 2 is development of cracks. Stage 3 is influx of groundwater.

a rake of 47.3°. The area and slip on the ruptured surface of the lower and upper fault-planes were used to calculate M_w (moment magnitude scale), with respective values of 6.7 and 6.3.

The total M_w was about 6.8 and agreed with the result of the moment tensor inversion solution from the Harvard CMT database (http://www.seismology.harvard.edu/), indicating that the coseismic energy were mainly released by the lower fault. Coseismic strain distribution due to the 2003 Chengkung earthquake was calculated using the dislocation fault model [17] and a computer code by [14]. Contraction surface strain near the Antung hot spring area was approximately 20 ppm (Fig. 6).

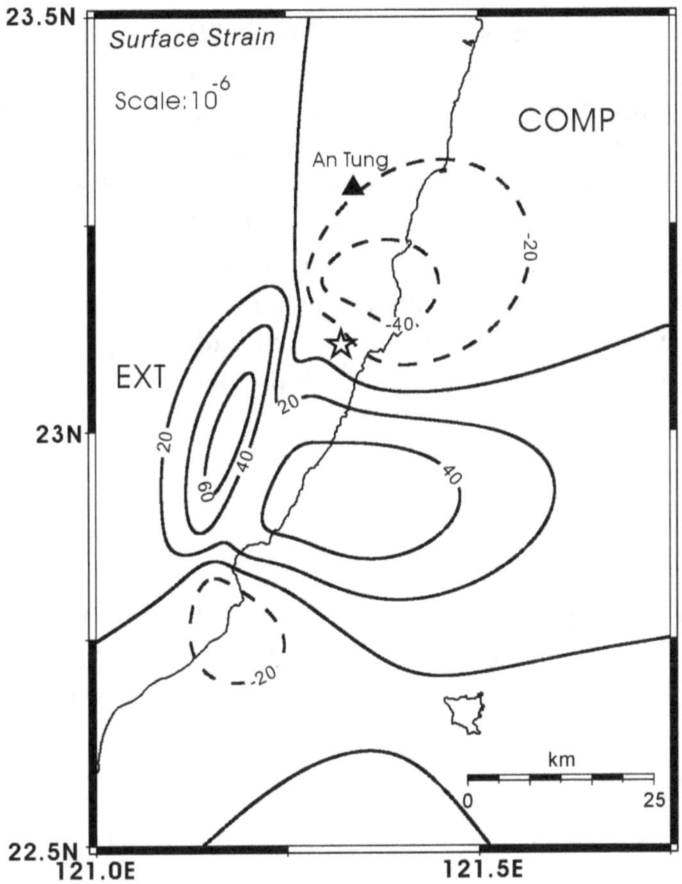

Figure 6. Distribution of coseismic surface strain (ppm) calculated based on the computer code for dislocation models by [14]. Positive and negative values mean dilatation and contraction, respectively. The open star denotes the 2003 mainshock. The filled triangle denotes the radon-monitoring well (D1). EXT and COMP denote dilatation and contraction, respectively.

A seismic slip can be preceded by accelerating aseismic slips near the hypocenter of an impending earthquake. Compared with the strain step about 20 ppm at the time of the 2003 Chengkung earthquake, the precursory strains from aseismic fault slips are small and accelerating. It is of practical importance to detect the accelerating transient of aseismic rock strain for the warning of local disastrous earthquakes. With the help of a case study, we show the capability to monitor the precursory decline in groundwater radon and to detect the accelerating transient of aseismic rock strain prior to the 2003 M_w = 6.8 Chengkung earthquake. Based on the precursory decline in groundwater radon observed at the Antung hot spring (Fig. 2), equation (3) can be used to calculate the accelerating transient of crustal-strain from aseismic fault slips prior to the 2003 M_w = 6.8 Chengkung earthquake. The open triangles in Fig. 2 show the calculated crustal-strain transient prior to 2003 M_W 6.8 Chengkung earthquake at the monitoring well (D1) in the Antung hot spring with an average fracture porosity of 0.00003.

5. Conclusions

In a series of laboratory experiments in un-drained conditions, I discovered a significant drop in groundwater radon, greater than 50 %, by a mechanism of radon volatilization into the gas phase. In-situ radon volatilization offers an attractive mechanism for a premonitory decrease in groundwater radon. An un-drained brittle aquifer near an active fault can be employed as a natural strain meter for detecting recurrent precursory radon declines. Anomalous declines in groundwater radon consistently recorded at a well (D1) prior to large earthquakes on the Chihshang fault in eastern Taiwan provide the reproducible evidence. Compared with the coseismic strain, the precursory strain from aseismic fault slips are transient and small. In this chapter, a quantitative method using the precursory radon decline as a tracer to detect the accelerating aseismic rock strain is presented with the help of a case study.

Acknowledgements

Supports by the National Science Council (NSC-103-2116-M-006-013), Central Geological Surveys, Industrial Technology Research Institute (L550001060), Radiation Monitoring Center, and Institute Earth Sciences of Academia Sinica of Taiwan are appreciated.

Author details

Ming-Ching T. Kuo

Address all correspondence to: mctkuobe@mail.ncku.edu.tw

Department of Mineral and Petroleum Engineering, National Cheng Kung University, Tainan, Taiwan

References

[1] Brace, W. F.; Paulding, B. W. Jr. & Scholz, C. (1966). Dilatancy in the fracture of crystalline rocks. Journal of Geophysical Research, Vol. 71, pp. 3939-3953, ISSN 2156–2202

[2] Central Geological Survey (2000a). Geological map of Taiwan in scale of 1 to 500000. Central Geological Survey, Taiwan

[3] Central Geological Survey (2000b). Active fault map of Taiwan in scale of 1 to 500000. Central Geological Survey, Taiwan

[4] Chen, W. S. & Wang, Y. (1996). Geology of the Coastal Range, eastern Taiwan, In: Geological Series of Taiwan 7, F. C. Chien, Central Geological Survey, ISBN 957-00-6978-3, Taiwan

[5] Clever, H. L. (1979). Krypton, Xenon and Radon-Gas Solubilities, In: Solubility Data Series 2, H. L. Clever & R. Battino, Pergamon Press, ISBN 0-08-022352-4, Oxford, UK

[6] Hsu, T. L. (1962). Recent faulting in the Longitudinal Valley of eastern Taiwan. Memoir of the Geological Society of China, Vol. 1, pp. 95-102, ISSN 0578-1825

[7] Kuo, M. C. Tom; Fan, K.; Kuochen, H. & Chen, W. (2006). A mechanism for anomalous decline in radon precursory to an earthquake. Ground Water, Vol. 44, pp. 642-647, ISSN 1745-6584

[8] Kuo, T.; Lin, C.; Chang, G.; Fan, K.; Cheng, W. & Lewis, C. (2010a). Estimation of aseimic crustal-strain using radon precursors of the 2003 M 6.8, 2006 M 6.1, and 2008 M 5.0 earthquakes in eastern Taiwan. Natural Hazards, Vol. 53, pp. 219-228, ISSN 1573-0840

[9] Kuo, T.; Cheng, W.; Lin, C.; Fan, K.; Chang, G. & Yang, T. (2010b). Simultaneous declines in radon and methane precursory to 2008 Mw 5.0 Antung earthquake: Corroboration of in-situ volatilization. Nat Hazards, Vol. 54, pp. 367-372, ISSN 1573-0840

[10] Kuo, T.; Liu, C.; Su, C., Chang, C.; Chen, W.; Chen, C.; Lin, C.; Kuochen, H.; Hsu, Y.; Lin, Y.; Huang, Y. & Lin, H. (2013). Concurrent concentration declines in groundwater-dissolved radon, methane and ethane precursory to 2011 Mw 5.0 Chimei eqrthquake. Radiation Measurements, Vol. 58, pp. 121-127, ISSN 1350-4487

[11] Kuo, T. (2014). Correlating precursory declines groundwater radon with earthquake magnitude. Groundwater, Vol. 52, pp. 217-224, ISSN 1745-6584

[12] Noguchi, M. (1964). Radioactivity measurement of radon by means of liquid scintillation fluid. Radioisotope, Vol. 13, pp. 362-366, ISSN 0033-8303

[13] Nur, A. (1972). Dilatancy, pore fluids, and premonitory variations of ts/tp traval times. Bulletin of the Seismological Society of America, Vol. 62, pp. 1217-1222, ISSN 1943-3573

[14] Okada, Y. (1992). Internal deformation due to shear and tensile faults in a half-space. Bulletin of the Seismological Society of America, Vol. 82, pp. 1018-1040, ISSN 0037-1106

[15] Scholz, C. H.; Sykes, L. R. & Aggarwal, Y. P. (1973). Earthquake prediction: A physical basis. Science, Vol. 181, pp. 803-810, ISSN 1095-9203

[16] Snow, D.T. (1968). Rock fracture spacings, openings, and porosities. Journal of the Soil Mechanics and Foundation Division, ASCE, Vol. 94, pp. 73-91, ISSN: 0044-7994

[17] Wu, Y. M.; Chen, Y. G.; Shin, T. C.; Kuochen, H.; Hou, C. S.; Hu, J. C.; Chang, C. H.; Wu, C. F. & Teng. T. L. (2006). Coseismic versus interseismic ground deformations, fault rupture inversion and segmentation revealed by 2003 Mw 6.8 Chengkung earthquake in eastern Taiwan. Geophysical Research Letter, Vol. 33, L02312, doi: 10.1029/2005GL024711, ISSN 0094-8276

6

Stability and Run-out Analysis of Earthquake-induced Landslides

Yingbin Zhang

1. Introduction

A large number of landslides can be caused by a strong earthquake and they have been the source of significant damage and loss of people and property. Therefore, it is very important to predict the stability of slope and the movement behaviors of a potential landslide under an earthquake loading, i.e., stability and run-out analysis (Figure 1).

Earthquake-induced landslides have been the source of significant damage and loss of people and property. One of the most serious event is the 1970 Peru earthquake. This event caused a huge rock avalanche that killed almost 54,000 people and buried two cities [143]. Another example is, in the 1920 Haiyuan earthquake, a large number of landslides caused widespread damage to infrastructure and buildings and killed at least 100,000 people, almost half of the total earthquake deaths [82].

Therefore, it is very important to predict the earthquake-induced landslides and to take countermeasures for potential landslides.

Main topics of earthquake-induced landslides are the following:

1. Investigation of recent and historical earthquake-induced landslides and their impacts so as to produce inventories of historical earthquake-induced landslides
2. Prediction of potential earthquake-induced landslides, including (i) failure mechanism and stability analysis of seismic slopes, (ii) movement mechanism and behaviours of earthquake-induced landslides, and (iii) Instrumentation and monitoring technologies for potential earthquake-induced landslides or post-earthquake landslides.
3. Preventive countermeasures for earthquake-induced landslides, including (i) Stabilization and disaster mitigation of earthquake-related landslides, (ii) risk assessment and

management of earthquake-related landslides, and (iii) hazard map and early warning system for earthquake-related landslides

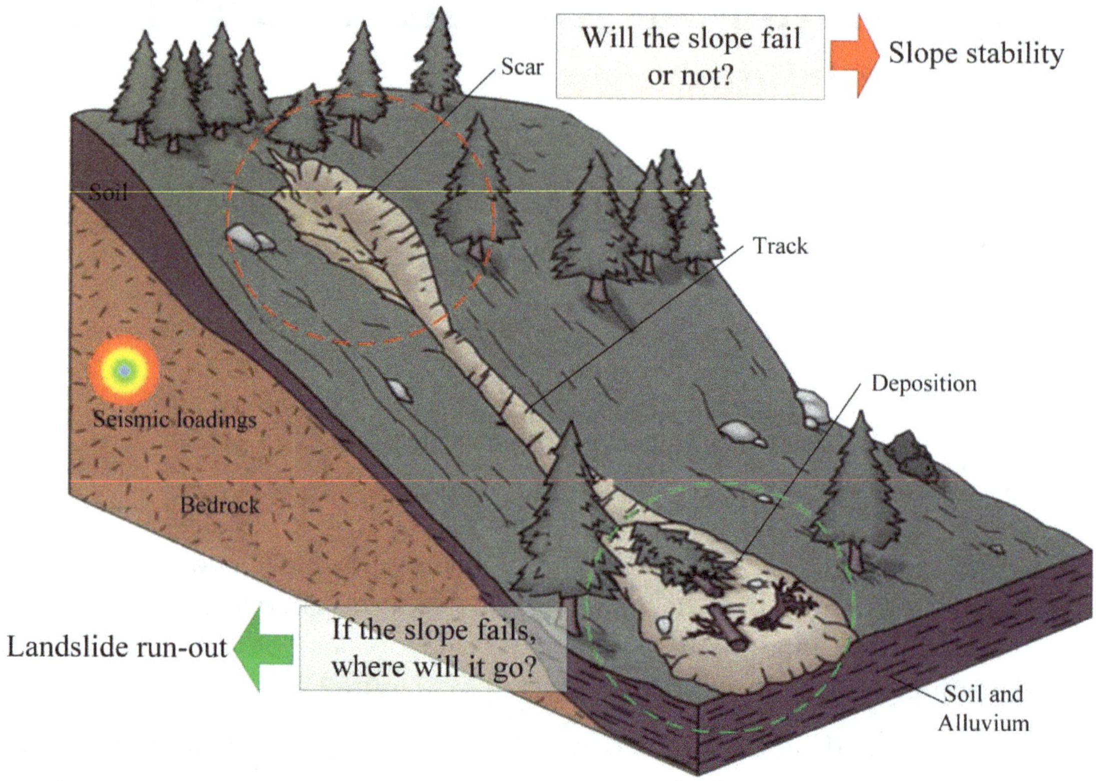

Figure 1. Stability and run-out analysis of earthquake-induced landslides

This chapter focuses on the prediction of potential earthquake-induced landslides. The prediction of potential landslide can be carried out using detailed geotechnical investigations and stability calculations. (i) Failure mechanism and stability analysis of seismic slopes, i.e. seismic slope stability analysis and (ii) movement mechanism and behaviours of earthquake-induced landslides, i.e. landside run-out analysis are outlined firstly, and then the merits and demerits of each method are clarified in this chapter.

2. Seismic slope stability analysis

So far, methods developed to analyze the stability of earthquake slopes can be divided into three types: (1) pseudo-static methods, (2) dynamic sliding block methods, and (3) stress-strain methods. These three types of methods can be applied in different cases due to each of them has merit and demerit [73].

2.1. Pseudo-static methods

[166] first presented the pseudo-static method, which is a simple method for evaluating of seismic stability of a slope. This type of method can be used to man-made or natural slopes

based on either analytical method or numerical method. The earthquake force, acting on the an element or whole of the slope, is writed by a horizontal force and/or a vertical volum force equal to the gravitation force multiple a coefficient *k*, called the pseudo-static coefficient as shown in Figure 2 and Equation (1).

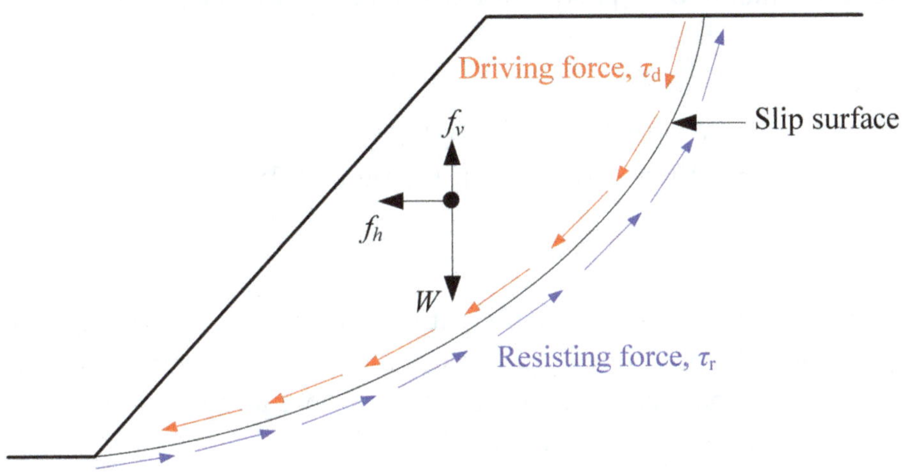

Figure 2. Forces acting on a slope in pseudo-static slope stability analysis

Thus, *k* times the gravitational acceleration *g*, i.e. *a*=*kg* forms the assumed seismic acceleration *a*. The assumed pseudo-static forces acting on a potential sliding mass of weight *W* will be

$$f_h = \frac{a_h W}{g} = k_h W$$
$$f_v = \frac{a_v W}{g} = k_v W \tag{1}$$

where a_h and a_v are horizontal and vertical pseudo-static accelerations, respectively, k_h and k_v are horizontal and vertical pseudo-static coefficients, respectively. The factor of safety(FOS) is represented as the ratio of the resisting force to the driving force, Equation (2).

$$\text{FOS}=\frac{\tau_r}{\tau_d} \tag{2}$$

From Equation (1), the pseudo-static force is determined by the seismic coefficient. The key problem for the pseudo-static procedure is how to select an appropriate seismic coefficient under an acceptable FOS. There have been studies for determining the most appropriate pseudo-static coefficient by a matter of experience and judgment.

[166] classical paper made the original suggestion to use of k_h=0.1 for severe earthquakes, k_h=0.2 for violent and/or destructive earthquakes, and of k_h=0.5 for catastrophic earthquakes.

[103] presented a minimum pseudo-static FOS of 1.5 based on a slope material strength reduction factor (SRF) of 0.8 and the following acceleration values associated with two different earthquake magnitudes *M*. The same values of seismic coefficients for magnitude 6.5 and 8.25 earthquakes are recommended by [154], but with an acceptable FOS of 1.15.

$$\begin{aligned} a &= 0.1g \text{ for } M = 6.5 \text{ implying } k = 0.1 \\ a &= 0.15g \text{ for } M = 8.25 \text{ implying } k = 0.15 \end{aligned} \tag{3}$$

[137] also presented the pseudo-static coefficient related to earthquake magnitude. In detail, for an 8.25, 7.5, 7.0 and 6.5 magnitude earthquakes, if the seismic coefficients equal to 1/2, 1/3, 1/4 and 1/5 of the PGA, respectively, the computed FOSs are larger than 1.0, the accumulated displacements of slope are likely to be acceptably small.

In the report published by the International Commission of Large Dams (ICOLD), [154] shows a list of the minimum FOS value and horizontal seismic coefficients for 14 large dams worldwide, in which the minimum FOSs range from 1.0 to 1.5 and the horizontal earthquake coefficients range from 0.1 to 0.15. The Corps of Engineers Manual recommended a earthquake coefficient of 0.1 or 0.15 for areas where major and great earthquake threats are estimated, respectively, and a FOS of no larger than 1.0 for all magnitude earthquakes.

Some references related the earthquake coefficient value to the peak ground acceleration (PGA) [10, 67, 108]. [108] related a pseudo-static coefficient of 1/3 to 1/2 of the PGA at the top of a double-side slope (a dam in the source reference), whereas [67] related a pseudo-static coefficient of 1/2 of the PGA of bedrock (PGA_{rock}) with a FOS of no larger than 1.0 and a SRF of 20%. And, [10] recommended the pseudo-static coefficient of 0.6 or 0.75 times of the PGA of bedrock (0.6 or $0.75PGA_{rock}$). It should be noted that the value given by [10] is conservative because the original study is designed for solid-waste landfills, where the allowable deformation are relatively small. [89] pointed that although engineering judgment is required for all cases, the criteria of [67] should be appropriate for most slopes.

[91] suggested one-half of PGA to use in an area of low seismicity (peak acceleration $<0.15g$) for the stability of earth embankments. This can be obtained from the peak horizontal motion (mean) from Modified Mercalli Intensity (MMI), magnitude-distance attenuation and the probability of a 50-year, 90% nonexceedance. However, in an area of moderate to strong seismicity ($0.15g \leq PGA \leq 0.40g$), PGA is obtained from the peak horizontal motion, from MMI, magnitude-distance attenuation and probability of 250-year, 90% nonexceedance.

[76] suggested a minimum FOS of 1.0, also based on a slope material SRF of 0.8 and the following values of pseudo-static coefficient: *a* equals to 0.17PGA or 0.5PGA for the dynamic response analysis is to be performed for the slope or earthquake structure or not.

3. Review of seismic zoning studies in Egypt

Seismic hazard assessments for Egypt, based on the zoning approach, has been carried out by many authors in the last decades, based upon the main tectonic features prevailed, the dominant tectonic stresses, the history of seismicity in the region, and the distribution of the recorded earthquakes. These authors were used different criteria to obtain seismic source zonation maps.

Among those studies, those carried out by the following authors: Sieberg [12 and 13], Gergawi and El-Khashab [15], Maamoun and Ibrahim [29], Maamoun *et al.* [16], Albert [30 and 31], Kebeasy *et al.* [32], Kebeasy [17 and 33], Marzouk [34], Fat-Helbary [35-37], Reborto *et al.* [38], Mohammed [39], El-Hadidy [40], Fat-Helbary and Ohta [41], El-Sayed and Wahlstörm [42], Abou Elenean [19 and 43], Badawy [44], Deif [45], Riad *et al.* [46], Abou Elenean and Deif [47], El-Sayed *et al.* [48], Fat-Helbary and Tealeb [49], El-Amin [50 and 51], El-Hefnawy *et al.* [52], Abdel-Rahman *et al.* [53], El-Hadidy [54 and 55], Deif *et al.* [56 and 57], Fat-Helbary *et al.* [58] and Mohamed *et al.* [59].

Egypt was divided into different seismic zones by many researchers, using the distribution of historical and instrumental earthquakes. Maamoun and Ibrahim [29] and Kebeasy [33] divided Egypt into four main seismic trends: i) Northern Red Sea-Gulf of Suez-Cairo-Alexandria, ii) Eastern Mediterranean-Cairo-Fayoum, iii) Mediterranean Coastal Dislocation, and iv) Aqaba-Dead Sea Transform. More recently, Maamoun *et al.* [16] added another two trends to the previous four: i) Hellenic and Cyprian Arcs, and ii) Southern Egyptian trend.

In reviewing the seismicity of Egypt, Kebeasy [17] suggested three main seismic zones: i) Aqaba-Dead Sea Transform, ii) Northern Red Sea-Gulf of Suez-Cairo-Alexandria, iii) Eastern Mediterranean-Cairo-Fayoum zones. In addition, he defined other local seismic zones (e.g., El-Gilf El-Kebeir, Aswan and Qena zones).

Fat-Helbary [36] assessed the seismic hazard for Aswan region. He used both of line sources and area source models. Five active faults in the Aswan region (Kalabsha, Seiyal, Gebel El-Barqa, Kurkur, and Khur El-Ramla Faults) were modeled as seismic lines. On the other hand, six area source zones (Old Stream, North Kalabsha, Khur El-Ramla, East Gebel Marawa, Abu Dirwa, and Kalabsha zones) were considered in the assessment. This study was followed by successive assessments by different authors to include other neighbor regions in Upper Egypt (e.g., [37, 41, 49, 50, 51, 57 and 58]).

Using the relation between the paleo-stresses, the present-day stresses and the distribution of earthquake epicenters, El-Hadidy [40] deduced five major trends in Egypt. They are: i) Pelusium megashear, ii) Eastern Mediterranean-Cairo-Fayoum-El-Gilf El-Kebeir, iii) Nubian-Mozambique, iv) Qena-Aqaba-Dead Sea, and v) Northern Red Sea-Gulf of Suez-Cairo-Alexandria seismotectonic trends. Furthermore, he identified some local zones on the Red Sea, Gulf of Suez, Gulf of Aqaba, Nile Delta, and Cairo-Suez regions.

According to the earthquake distribution, focal mechanisms and the structural and tectonic information, Abou Elenean [19] suggested five seismotectonic sources. They are: i) Gulf of

Suez-Northern Eastern Desert, ii) Southwest Cairo (Dahshour), iii) Northern Red Sea, iv) Gulf of Aqaba, and v) Aswan zones. Deif [45], for a seismic hazard assessment study, delineated four additional seismic sources for the southern part of Egypt. They are: i) Abu Dabbab, ii) El-Gilf El-Kebeir, iii) Wadi Halfa, and iv) Northern Nasser's Lake zones.

Riad *et al.* [46] constructed a more detailed seismic zoning map for Egypt and its surroundings. Their regional delineation consists of five main trends: i) the Greek trend, based on the seismic zone regionalization of Papazachos [61], ii) the Dead Sea trend, which mainly based on the earthquake catalogue of Israel and its vicinity [62], iii) Pelusium and Qattara trend, iv) Eastern Mediterranean trend, and v) Aswan area, in Southern Egypt.

El-Hefnawy *et al.* [52], based on the tectonic regime, seismicity, faults location, and focal mechanism solutions, divided the regional seismicity in and around Sinai Peninsula into 25 source zones. His study was succeeded by a certain number of studies that considered a more detailed zonation for the same area (e.g., [53, 54 and 56]).

Recently, Abou Elenean [43] established a detailed zonation map for whole Egypt and its surroundings, considering the recent seismicity distribution and focal mechanism data. He delineated 41 seismic source zones of shallow-depth earthquakes (h < 60 km) in and around Egypt. In addition, he considered 7 seismic sources for intermediate-depth events within the Hellenic Arc (after [63]). More recently, El-Hadidy [55] and Mohamed *et al.* [59] established a new and modified seismic zoning map for Egypt and its surroundings which is based on the compilation of previous studies [53, 57 and 64].

4. Data sources

For the construction of any database of seismic sources, there are two basic steps: first, all of the active faults that affect a specific region need to be recognized, and secondly, each seismogenic structure should be seismotectonically parameterized. In order to recognize the active faults, it is necessary to analyze the seismicity. It is common practice to start analyzing the historical and instrumental seismicity that affects the specific region. Like many other places all over the world, the seismicity in Egypt is not homogeneously distributed, neither in frequency nor in density. Historical information is similarly not uniform all over the region.

4.1. An updated earthquake catalogue

A complete and consistent earthquake catalogue in a region is essential in order to study the distribution of earthquakes in space, time, and magnitude. In the current work, the identification and characterization of regional seismic source zones is based on a unified compiled earthquake catalogue, after Sawires *et al.* [60], for Egypt and its surroundings which covers the area from 21° to 38° N and 22° to 38° E, and extends from 2200 B.C. until 2013 in the time period.

Different earthquake magnitude scaling relations, correlating different scale magnitudes, were used to develop a unified earthquake catalogue for the study region in the moment magnitude

(M_W) scale. The dependent events were removed from the catalogue to ensure a time-independent (Poissonian) distribution of earthquakes (Figure 3).

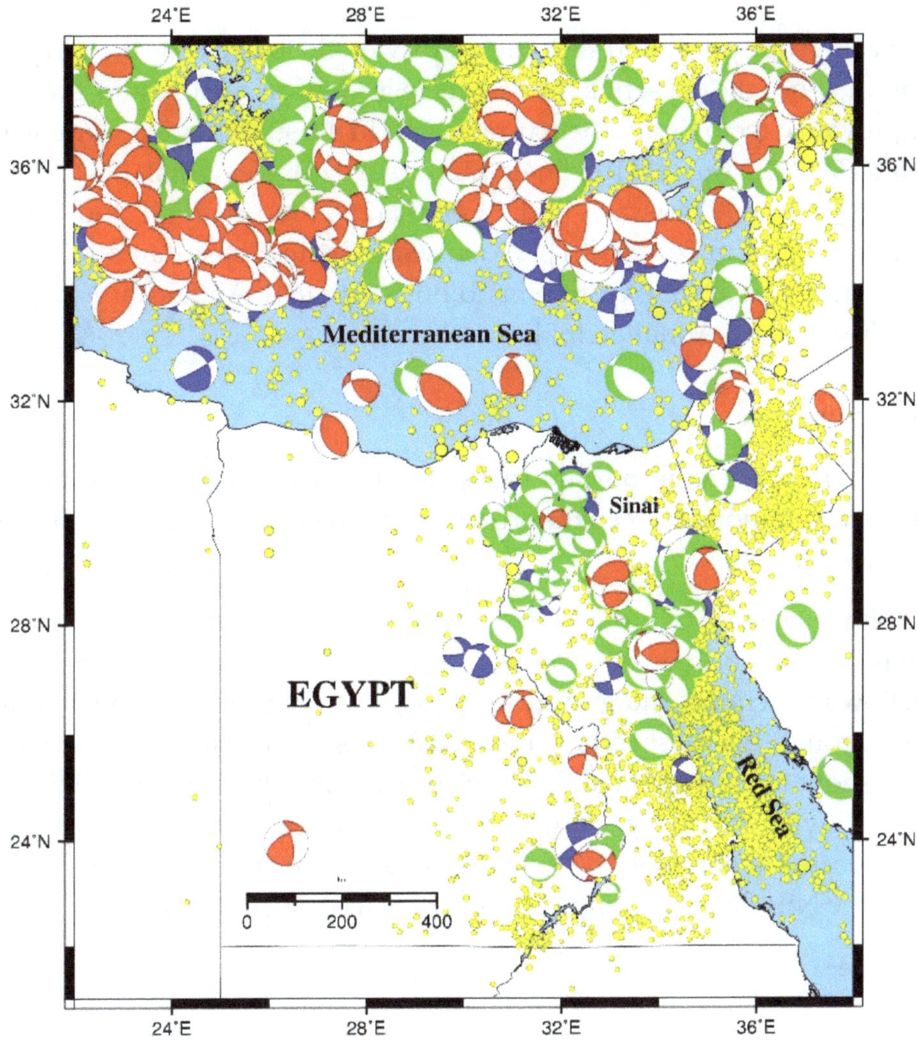

Figure 3. Distribution of the seismicity (2200 B.C. - 2013) and focal mechanism solutions (1940 – 2013) in and around Egypt (after Sawires *et al.* [60]). Symbols and focal sphere sizes are in proportion the moment magnitude. Focal sphere colours refer to different fault types (blue: strike-slip; green: normal; red: reverse).

4.2. Focal mechanism data

Different local and international sources were examined and focal mechanism data were compiled into a single database. The solutions of the Global Catalogue of CMT Harvard [65], the International Seismological Centre (ISC) [66], the National Earthquake Information Centre (NEIC) [67], the Regional CMT catalogues (RCMT) in the Mediterranean region [68], as well as ZUR-RMT catalogue of the Institute of Technology (ETH) of Zurich were also included in the catalogue. More than 600 focal mechanism solutions were collected covering different active seismic zones (Figure 3) in Egypt and surroundings, spanning the spatial area from 21°

to 38°N, and from 22° to 38°E. Most of them have a magnitude greater than or equal to M_W 3.0, occurring in the time period 1940 to 2013.

4.3. Geological, tectonic and geophysical data

Several geological, geophysical and tectonic maps were inspected for the purpose of getting more information about the present active faults (e.g., Aswan region) and also for the identification of the prevailed tectonic and structural trends in the study region. Among these studies are those of Said [69-71], Youssef [10], Shata [72], Neev [73], Neev *et al.* [74 and 75], El-Shazly [76], Riad [27], Maamoun [77], Issawi [78], EGSMA [26], Riad *et al.* [47 and 79], Maamoun *et al.* [16], Sestini [80], Schlumberger [81], Woodward-Clyde Consultants [82], Kebeasy [17], Meshref [11], Barazangi *et al.* [83], Guiraud and Bosworth [84], Abdel Aal *et al.* [85], Philobbos *et al.* [86], and Hussein and Abdallah [87].

4.4. Crustal structure data

The crustal structure plays an important role in Seismology. It can be used, as in the current study, for the discrimination between the crustal (shallow-depth) seismicity, the intermediate-depth, and the deeper one.

Several studies have been carried out to evaluate the crustal structure and thickness in Egypt by using different types of datasets coming from seismic reflection surveys, deep seismic sounding, shallow refractions, and gravity (e.g., [34, 40 and 88-108]). In the delimitation of the different seismic zones, the most recent study [108] was taken into our consideration (Figure 4). Their results show that the Moho discontinuity is getting shallow toward the northern and eastern coast of Egypt, and deeper toward Western Desert and Northeastern Sinai. This discontinuity is located at depth of 31-33 km in Greater Cairo and Dahshour, 32-35 km in Sinai, 33–35 km along the Nile River, 30 km near the Red Sea coast, and 39 km towards the Western Desert.

5. Detailed description of the new proposed shallow-depth seismic source model

Seismic sources define areas that share common seismological, tectonic, and geologic attributes, and that can be described by a unique magnitude-frequency relation. In terms of PSHA, a seismic source represents a region of the earth's crust in which future seismicity is assumed to follow specified probability distributions for occurrences in time, earthquake sizes, and locations in space [109].

Araya and Der Kiureghian [109] discriminate between seismogenic and seismicity sources. Seismogenic zones lack the development of a clear history relating the contemporary seismic activity to a geologic structure. For such zones, critical gaps in the Quaternary geologic history preclude direct evidence of active faulting. Seismogenic zones are, by far, the most common type of source zone employed in PSHA. Commonly, seismogenic zones are area sources, but the zone type applies also to inferred associations of seismicity with individual faults. On the other hand, seismicity zones are source zones that are defined with no consideration of their relation to geologic structures. They are defined solely based on the spatial distributions of the seismic history, and their use and reasonableness can only be judged relative to the intended use of the final hazard estimate. This will be the terminology used in this work.

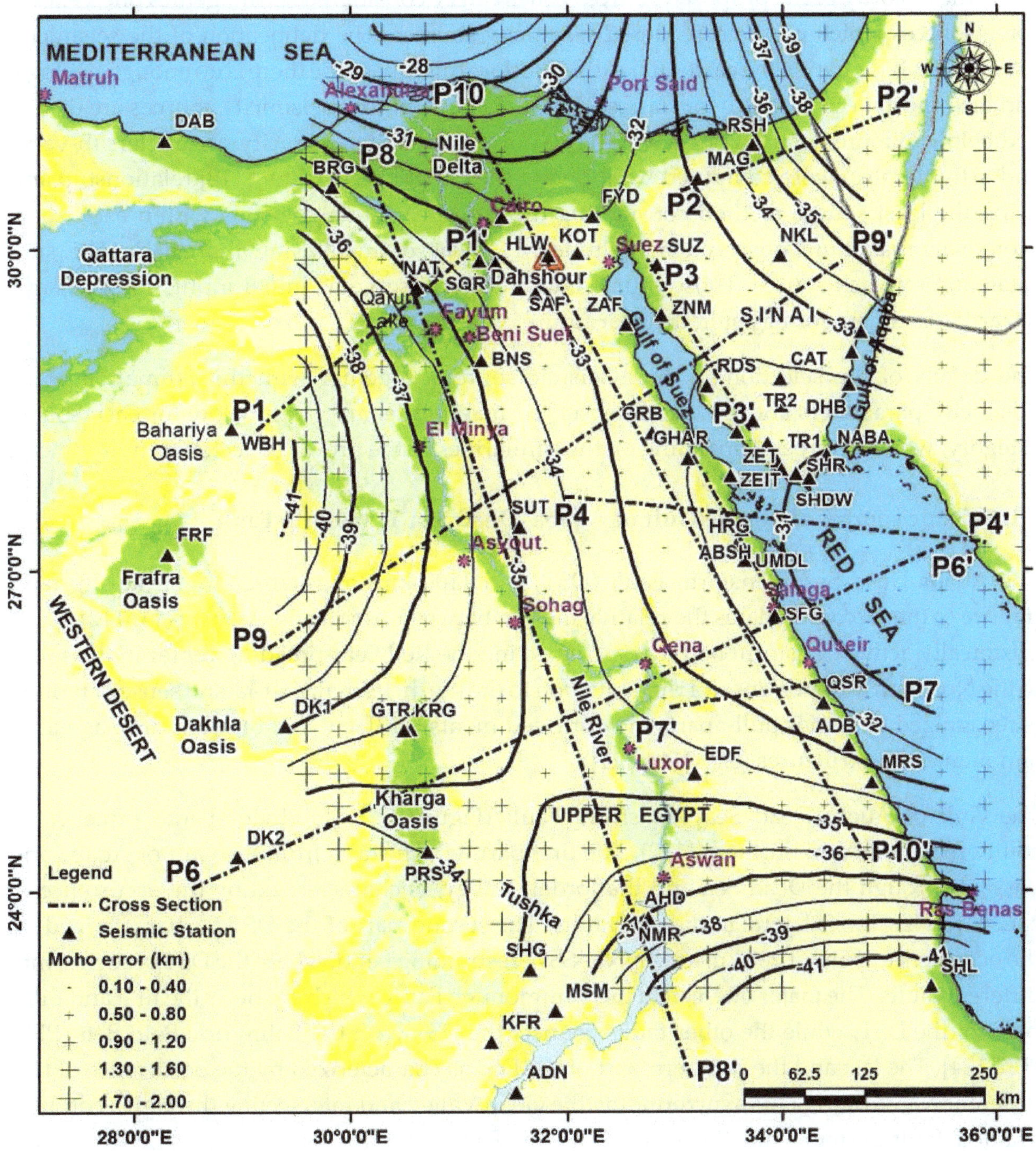

Figure 4. Depth of Moho discontinuity in Egypt (after Abdelwahed *et al.* [108]).

As mentioned previously, the separation of the study area into smaller, seismotectonically homogeneous zones is based on criteria mainly related with the present-day tectonic regime, epicenter distribution, focal mechanism data and the location of known faults. In the present work, we decided to employ simple geometric shapes for the definition of the seismic source model. The regional seismicity of concern to Egypt was divided into 28 seismic sources (Figure 5). These zones was related to the tectonic activity of the previously defined local active belts. Thus, the majority of the proposed sources zones can be considered seismogenic zones, except some sources which can be considered seismicity sources. The delineation of the seismicity sources was based upon the earthquake distribution, this is because there is no enough geologic and tectonic data covering these sources. Both seismogenic and seismicity sources are described below in more details. For each of these source zones, the seismicity parameters (b-value and activity rates) were computed by applying the Gutenberg-Richter [110] relationship and using the least square method considering the entire earthquake events within each zone. Moreover, maximum observed magnitude M_{max} was defined using the earthquake sub-catalogue for each source. Those estimated values will serve as initial inputs for a seismic hazard assessment for Egypt in the near future.

The details of the selection of these seismic sources, together with the estimation of its seismicity parameters and maximum observed magnitude, are given below for each source category, which grouped depending on the similarities of the prevailed tectonic environment.

5.1. Seismic sources along the Gulf of Aqaba–Dead Sea Transform Fault

The Aqaba-Dead Sea Transform Fault (DST) is a 1100 km long left-lateral strike-slip fault (Figure 6) that accommodates the relative motion between Africa and Arabia [111, 112]. It is a seismically active transform boundary, connecting the Red Sea spreading center in the south to the Northern Mediterranean Triple Junction to the north. Its main left-lateral sense of motion is recognized by minor pull-aparts in young sediments [113], cut and offset of drainage lines and man-made structures (e.g., [114-121]).

The Gulf of Aqaba-Dead Sea Transform Fault (Figure 6) is subdivided into three parts; southern, central and northern [122]. The first part, which starts from the Gulf of Aqaba and passing through the Dead Sea and the Jordan Valley, is characterized by the occurrence of N12°E to N20°E left-lateral strike-slip faults. The second part of the DST is characterized by the occurrence of about 200 km long NNE–SSW restraining bend, where the DST branches into different faults. The major one, called the Yammouneh Fault, which connects the first and third parts of the DST, while the other faults connect the DST with the Palmyride Fold Belt (PFB) [122-124]. The last and the northern part of the DST is characterized by the occurrence of two different N–S striking faults surrounding the Ghab Valley and intersecting through a complex braided fault system with the East Anatolian Fault and the Cyprian Arc [125-127]. This intersection corresponds to the Hatay "fault–fault–trench" triple junction that forms the plate boundaries between Arabia, Africa and Anatolia [128].

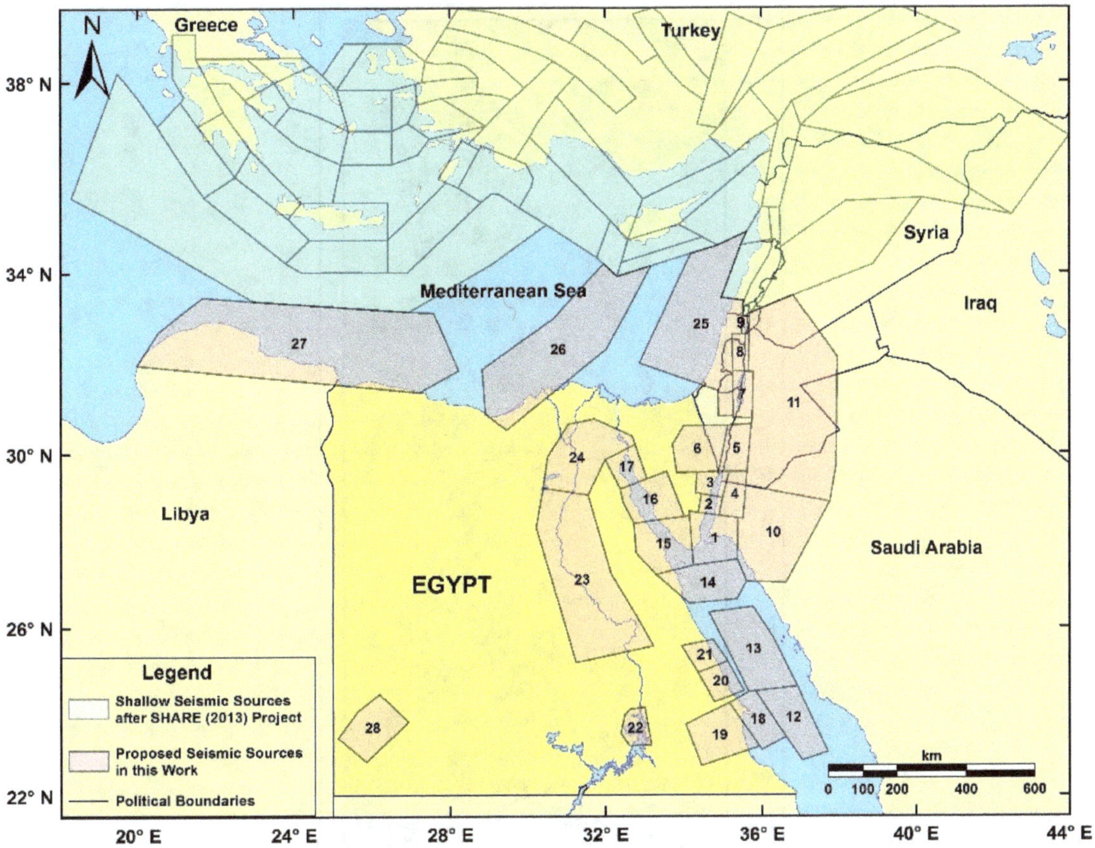

Figure 5. Proposed seismic source zones in Egypt and its surroundings.

5.1.1. Gulf of Aqaba seismogenic sources (EG-01 till EG-04)

The Gulf of Aqaba experienced the largest Egyptian earthquake (M_W 7.2, November 1995) which struck the area and its effects were extending till Cairo. Over than 1000 aftershocks are recorded. The aftershocks area reached a length of about 110 km, striking N 30° E, which in turn parallel to the Gulf of Aqaba trend [129]. Potential damage was observed at Nuweiba city at the western part of the gulf.

The Gulf of Aqaba has been considered to be the most active seismic area over the last few decades, characterized by swarm activity [130-132]. There is no information about the seismicity of the Gulf of Aqaba until the year 1983. However, from January till April 1983, over than 500 events were reported, reaching a maximum recorded magnitude of 4.8. These earthquake events were felt at different places along the gulf area, as well as along the Arava Valley founding a general consideration [133]. From August 1993 up to February 1994, a large earthquake swarm was associated with relatively high magnitudes, reaching a 5.8 value. This swarm included about 1200 events occurred south to the 1983 swarm. Another earthquake swarm has been recorded and located at the central part of the Gulf of Aqaba on November 2002. Over than 10 events with magnitude above 4.0 were recognized, and many other events with magnitudes below this value. Some of these earthquakes were felt, but without damage for buildings at the epicentral area.

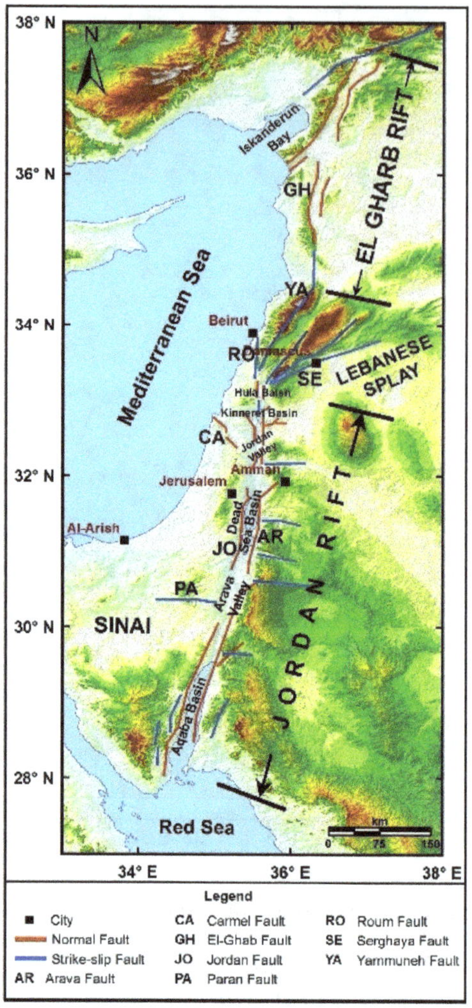

Figure 6. The main structural elements along the DST (redrawn after Heidbach and Ben-Avraham [134]).

The interior of the Gulf of Aqaba is occupied by three elongated en-echelon basins transected by longitudinal faults [131]. This en-echelon system produces several tectonic basins, which are forming rhombic-shaped grabens. Thus, three basins in the Gulf of Aqaba are present. They are, from south to north, Tiran "Arnona"-Dakar, Aragonese and Elat "Aqaba" Basins.

The heterogeneity of the focal mechanism solutions for the earthquake events taken place in the gulf area, indicates its geologic structure complexity. Some fault plane solutions exhibit normal faulting, which are related to the faults that form the boundaries of the major basins in the gulf. Others indicate left-lateral motion of the transform [112]. The focal mechanism of the M_W 7.2, 1995 Aqaba earthquake as well as some aftershocks, show a strike-slip movement with predominant normal components, with the exception of only one solution located on the eastern coast of the Gulf of Aqaba, and exhibits strike-slip movement with a little reverse component in the NNW-SSE and ENE-WNW nodal planes [19].

According to the seismic activity, the epicentral distribution and the local tectonics, different seismogenic sources were delineated in the gulf area (Figure 7).

a. The EG-01 (Tiran – Dakar Basin) seismogenic source lies at the southern part of the Gulf of Aqaba. It includes the M_S 4.4, February 2, 2006 earthquake. There is no historical earthquakes included in this source zone. The majority of the available focal mechanism solutions inside this area source reflects normal faulting mechanism.

b. The EG-02 (Aragonese Basin) seismogenic source lies to the north of the previous EG-01 zone, and is considered the focal area of the M_W 7.2, November 22, 1995 earthquake, which is considered the largest event to occur along the DST in the last century.

c. The EG-03 (Elat Basin) seismogenic source located to the north of the EG-02 seismic zone and considered as the extension area of the M_W 7.2, 1995 Aqaba earthquake rupture. It is characterized by a low seismicity level, if compared with the other two zones of the Gulf of Aqaba. Two historical events have been included in this area source, the I_{max} VIII, March 18, 1068, and the I_{max} VIII-IX, May 2, 1212 earthquakes.

d. In addition to the previous seismogenic sources, a delineation of a separate and fourth zone is taken place. This source lies to the east of the gulf and characterized by dispersed moderate seismicity. This zone is the EG-04 (Eastern Gulf of Aqaba) seismogenic source. The major earthquake included in this area source is the m_b 4.5 December 26, 1995 earthquake.

Previous focal mechanism solution studies for moderate to large earthquakes located in the Gulf of Aqaba region (e.g., [135-138]) assert the dominance of ENE-WSW extension (N60°-80°E). Furthermore, field studies [139, 140] observed two conjugate faults along the Gulf of Aqaba: NNE left-lateral strike-slip faults parallel to the gulf that release the majority of stress, and a nearly ESE-WNW normal faults along the margins of pull-apart basins. On the other hand, body waveform inversion of the M_W 6.1, August 3, 1993, and the M_W 7.2 November 22, 1995 events, support the occurrence of normal faulting take place along the transverse NNW-SSE and ESE-WNW faults, while left-lateral strike-slip movement occurs along NNE major Aqaba trend [135].

5.1.2. Arava Valley (EG-05) seismogenic source

The Arava Valley is located to the north of the Gulf of Aqaba. It is an inter-basin zone trending NE-SW. Its faults extend over 160 km from the Gulf of Aqaba to the Dead Sea and provide morphological evidence of essentially strike-slip motion [120]. It is characterized by a low seismicity level compared with the surrounding area, despite clear indications of recent faulting [141]. Klinger *et al.* [120] emphasized the limited earthquake activity in the Arava Valley in the instrumental period. Shapira and Jarradat [133] stated that, from preliminary paleoseismicity studies, the border-faults of Arava Valley generate earthquakes bigger than magnitude 6.0 with an average return period of 1000-3000 years.

There is no historical earthquakes included in this seismogenic source zone. The biggest recorded event is the m_b 5.2, December 18, 1956 earthquake. Two focal mechanism solutions are known in the northern part of this source, both of them exhibiting strike-slip faulting with normal component.

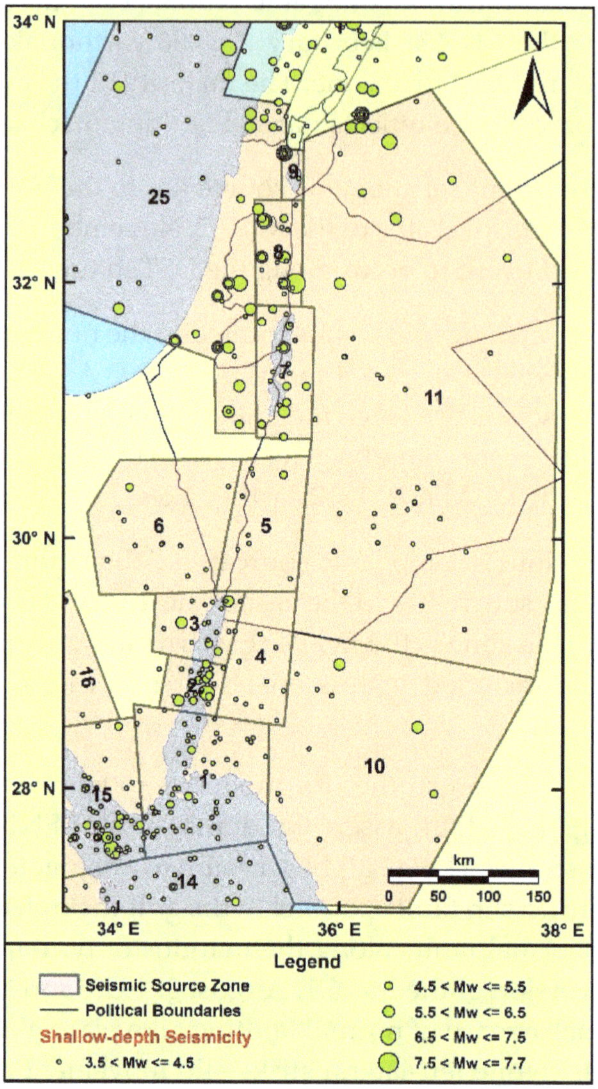

Figure 7. Shallow-depth seismicity (h ≤ 35 km) and delineated seismic sources along the Gulf of Aqaba-Dead Sea Transform Fault.

5.1.3. Eastern Central Sinai (EG-06) seismogenic source

An E-W trending dextral strike-slip faults with up to 2.5 km of displacement has been recognized in central Sinai by Steintz *et al.* [142]. It is called the Themed Fault. The Tih Plateau (in central Sinai) is traversed by the Themed Fault, which extends for about 200 km from the vicinity of eastern margin of the Suez Rift to the DST [71]. The Themed Fault has been reactivated along a pre-existing fault, identifying the southern border of the Early Mesozoic passive continental margin of the Eastern Mediterranean Basin in central Sinai [143].

To the north of the previous fault, the central Sinai-Negav shear zone is located, which is proposed by Shata [72] and Bartov [144]. It is a narrow E to ENE trending fault belt discriminating and separating the North Sinai Fold Belt (tectonically unstable area) from the Tih Plateau (tectonically stable area) in middle and Southern Sinai [145].

The EG-06 seismogenic source lies to the west of the previous EG-05 source and to the east of the Sinai sub-Plate. This seismogenic source includes the low seismic activity related to the Themed Fault, central Sinai-Negav shear zone, Paran Fault and Baraq/Paran Fault junction. This source has a great tectonic effect on Sinai Peninsula and its surrounding areas. There is no historical earthquakes included in this source, and the biggest earthquake located in this zone is the m_b 4.8, September 24, 1927 event.

5.1.4. Dead Sea Basin (EG-07) seismogenic source

The Dead Sea Basin is characterized by a double fault system that is bounded by the Arava Fault from the east, and by the Jordan (Jericho) Fault from the west, hence it occupies a rhomb-shaped graben between two left-lateral slip faults. The average slip rate on the Dead Sea portion of the transform fault is estimated to be 0.7 cm/yr. [114], which is consistent with the average slip of the overall plate boundary of 0.7-1.0 cm/yr.

Earthquake swarms and a mainshock-aftershock type of activity characterize this seismogenic source. Trenching studies across the Jordan Fault indicate that two large earthquake swarms occurred since about 2000 years ago. One of them is between 200 B.C.- 200 A.D., while the other one is between 700 A.D.- 900 A.D. [114]. El-Isa *et al.* [146] attributes these swarms to subsurface magmatic activity and/or to the isostatic adjustments along the Gulf of Aqaba.

Several historical earthquakes are included inside this source zone. They are the 745 B.C., 33 A.D., 1048, 1212, 1293, and 1458 earthquakes. Their intensities range between VII to VIII. Ben-Menahem *et al.* [147] obtained focal mechanism solutions for some recent events (e.g., the M_S 4.9, October 8, 1970 earthquake) which took place in the Dead Sea area. All solutions indicate a left-lateral strike-slip movement on a sub-vertical fault striking with an average trend of N8°-10°E. However, Salamon *et al.* [112] obtained normal focal mechanism solutions for some relatively recent events. These solutions may be describe the earthquake activity of the N-S striking normal faults bordering the Dead Sea Basin. Field observations confirmed this type of activity [113, 148].

5.1.5. Jordan Valley (EG-08) seismogenic source

The Jordan Valley trends in the N-S direction, linking between the Hula Basin to the north and the Dead Sea Basin to the south. The details about its end in the Sea of Galilee are not clear from the surface features [147]. Garfunkel *et al.* [113] noticed a small amount of compression along the valley and near the Jordan Fault trace. Recent earthquake activity along the Jordan Valley is low compared to the Southern Dead Sea Basin. Ten historical events (before 1900) are included in this area source. They are the 1020 B.C., 578 A.D., 580, 746, 854, 1034, 1105, 1160, 1260, and 1287 events. Their intensities range between IV to XI. The most important earthquake included in this source zone is the I_{max} XI, 746 event.

5.1.6. Kineret-Hula Basin (EG-09) seismogenic source

To the north of the previous Jordan Valley source are located the Hula (Shamir-Almagor Fault) and Kineret (Kineret-Sheikh Ali Fault) Basins [149]. Seismic activity in the two mentioned basins was located till the Yammuneh Fault (NE-bend of the Dead Sea Transform). This area source, which surrounded by the Roum Fault from the western side and the Jordan Fault from the eastern side, was considered by Shamir *et al.* [150] as a seismogenic step zone. Three historical events are included in this zone. They are the 19 A.D., 419, and 756 earthquakes. The biggest earthquake is the I_{max} X, 19 A.D. event.

5.1.7. Northwestern Saudi Arabia (EG-10) seismicity source

To the east of the EG-01 and EG-02 seismogenic sources, the Northwestern Saudi Arabia EG-10 source has been considered. This source zone covers disperse, low seismicity in the north-western part of Saudi Arabia. Two historical events are reported to occur inside this area source. They are the March 18, 1068, and January 4, 1588 earthquakes, both of them with intensity VIII.

5.1.8. Lebanon (EG-11) seismicity source

To the north of the previous seismic source and along the eastern boundaries of the EG-04, EG-05, EG-07, EG-08, and EG-09 sources, the Lebanon EG-11 seismicity source has been considered. This area source covers a dense disperse low-magnitude seismicity in Lebanon and Southern Syria. Nine historical earthquakes are located inside this area source. The most important among them are the 972, 1159, and 1182 events. Their felt intensities are I_{max} IX, IX-X, and IX, respectively.

The computed b-value, the annual rate of earthquakes, and the observed recorded maximum magnitude for the delineated seismic sources along the Gulf of Aqaba-Dead Sea Transform Fault are displayed in Table 1.

Source Zone	b-value	Yearly Number of Earthquakes		Observed M_{max}
		Above M_W 4.0	Above M_W 5.0	
EG-01	1.13	0.9799	0.0732	m_b 4.4 on 2006/02/02
EG-02	0.98	0.4952	0.0521	M_W 7.2 on 1995/11/22
EG-03	0.97	0.2763	0.0296	I_{max} VIII-IX on 1212/05/02
EG-04	1.01	0.1961	0.0191	m_b 4.5 on 1995/12/26
EG-05	0.88	0.1882	0.0251	m_b 5.2 on 1956/12/18
EG-06	1.12	0.1853	0.0140	m_b 4.8 on 1927/09/24
EG-07	0.87	0.3232	0.0438	I_{max} VIII on 1458/11/12*
EG-08	0.71	0.1865	0.0366	I_{max} XI on 0746/--/--
EG-09	0.91	0.0651	0.0080	I_{max} X on 0019/--/--
EG-10	1.03	0.1934	0.0180	I_{max} VIII on 1588/01/04*
EG-11	0.97	0.3645	0.0388	I_{max} IX-X on 1159/06/06

(*) the most recent event.

Table 1. b-value, annual rate of earthquakes, and maximum observed magnitude for the delineated seismic source zones along the Gulf of Aqaba-Dead Sea Transform Fault.

5.2. Seismic sources along the Red Sea Rift

The Arabian Plate is continuing to rotate away from the African Plate along the Red Sea Rift spreading center. The Red Sea occupies a long and slightly sinuous NW-trending escarpment-bound basin, 250-450 km wide and 1900 km long, between the uplifted shoulders of the African and Arabian shields. It is part of a rift system extending from the Gulf of Aden to the northern end of the Gulf of Suez. The overall trend of the rift is N30°W, although a few kinks occur at around 15°N, 18°N, and 22°N.

Depending on the structural setting and morphology of the Red Sea, it can be subdivided into four different zones (Figure 8). Each zone are representing distinct stage in the development of the continental margin and the generation of the mid-ocean ridge spreading system [151, 152]. These zones are:

i. *Active sea-floor spreading (Southern Red Sea):* It is located between 15°N and 20°N and characterized by a well-developed axial trough which has developed through normal sea-floor spreading during the last 5 Ma [153-155] or even older, at about 9–12 Ma [156].

ii. *Transition zone (central Red Sea):* It is located between 20°N to about 23°20′N, where the axial trough becomes discontinuous, in which the central Red Sea consists of a series of 'deeps' alternating with shallow 'inter-trough zones' [157]. An identical zone may flanks the deep axial trough between the side walls of the shallow main trough on both sides of other zones [151, 152].

iii. *Late stage continental rifting (Northern Red Sea):* This zone composed of a wide trough without a distinct spreading center, in spite of a number of small isolated "deeps" is occurred [152].

iv. *Active rifting:* This zone representing the expected line along which the Southern Red Sea may be propagate through the Danakil Depression Afar. This zone may be considered separately or it can be added to the first mentioned zone.

Based on the morphological and structural features of the Red Sea, the Egyptian part (northern latitude 22°N) can be divided into three distinct seismogenic source zones (EG-12, EG-13, and EG-14) (Figure 9). Each zone represents different stage of development [159]. The delineation is made, based upon the occurrence of the transverse structures, change of the fault trend along the axial rift and the variety of the seismic activity along the rift axis.

5.2.1. Southern Red Sea (EG-12) seismogenic source

The EG-12 Southern Egyptian Red Sea seismogenic source represents the northern part of the transition zone. It is characterized by NW-SE trending faults. The boundary proposed by Bonati [160], north latitude 25ºN, is found herein to coincide with the NE-trending transform faults and the associated seismicity. Only one historical event is included in this seismic source, the I_{max} VI-VII, 1121 earthquake.

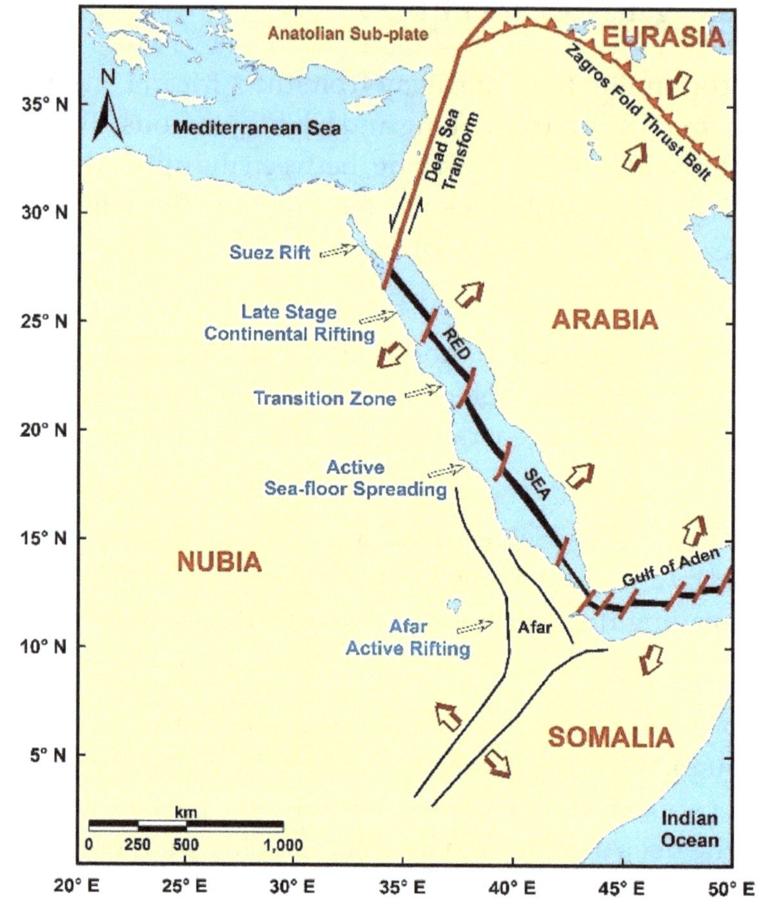

Figure 8. Tectonic framework of the Red Sea region (redrawn after Ghebreab [158]).

5.2.2. Central Red Sea (EG-13) seismogenic source

The EG-13 Central Egyptian Red Sea seismogenic source is located to the northwest of the previous zone. It corresponds to the region north of latitude 24º30′N, which consists of a broad main trough without a recognizable spreading center [152]. Recent recorded seismicity could indicate the expected location of the axial rift. In this zone, the degree of seismicity is relatively low and scattered, compared to the previous zone. Like the previous zone, there is only one historical event included here. It is the I_{max} V, 1899 earthquake. The maximum observed magnitude along this source corresponds to the m_b 4.7 (M_S 5.1), July 30, 2006 earthquake.

5.2.3. Northern Red Sea (EG-14) seismogenic source

The EG-14 Northern Egyptian Red Sea seismogenic source is characterized by higher seismic activity than the previous two sources. This activity may be due to the juncture between the two gulfs. Daggett *et al.* [161] studies of the low-magnitude seismicity shows that, the high seismic activity of the northern Red Sea is different from the activity at the southern part of the Gulf of Suez. There is no earthquakes related to this area source before the year 1900. In addition, the m_b 5.0 (M_S 5.0) March 22, 1952 event represents the biggest recorded earthquake till now.

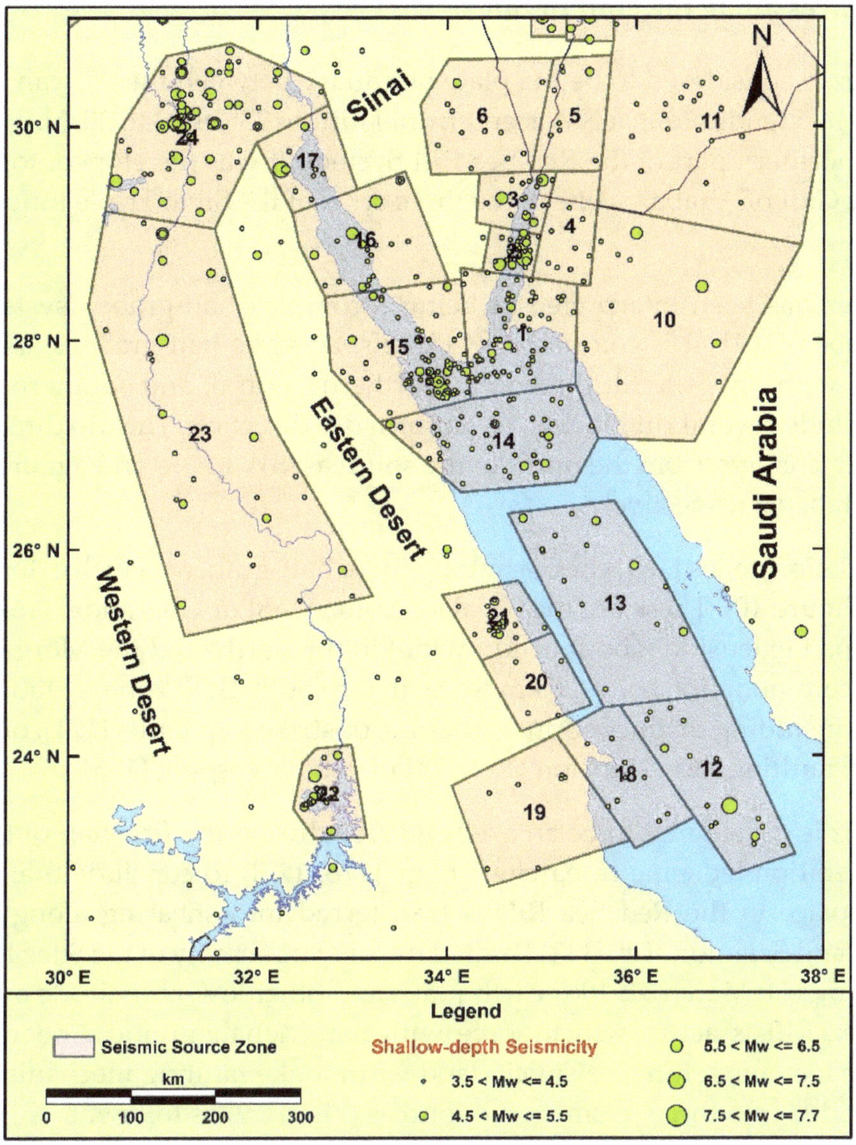

Figure 9. Shallow-depth seismicity (h ≤ 35 km) and delineated seismic sources along the Red Sea-Gulf of Suez and the Nile River.

Seismicity parameters for the delineated seismic sources along the Red Sea Rift are displayed in Table 2.

Source Zone	b-value	Yearly Number of Earthquakes		Observed M_{max}
		Above M_W 4.0	Above M_W 5.0	
EG-12	1.00	0.4359	0.0434	I_{max} VI-VII on 1121/--/--
EG-13	0.91	0.3029	0.0376	m_b 4.7 on 2006/07/30
EG-14	1.13	0.6425	0.0472	m_b 5.0 on 1952/03/22

Table 2. b-value, annual rate of earthquakes, and maximum observed magnitude for the delineated seismic source zones along the Red Sea Rift.

5.3. Seismic sources along the Gulf of Suez

The Gulf of Suez is considered to be the plate boundary between the African Plate and Sinai sub-Plate [162]. It extends along a NW trend from latitude 27°30′ N to 30°N. The Gulf of Suez constitutes the northern part of the Red Sea Rift System. It was developed, together with the Red Sea and the Gulf of Aqaba, as one of the three arms of the Sinai Triple Junction [69, 81 and 163-166].

The Gulf of Suez has been interpreted as being a complex half-graben system [139], or an asymmetric graben [167]. It is composed of three successive half-grabens, as mentioned by Moustafa [168], with opposite tilt directions: northern, central, and southern. These distinct half-grabens include several rift blocks of a uniform dip direction. The dip direction, along the Gulf of Suez Rift, changes from the north to the south as: SW to NE and again to SW defining the three half-grabens, respectively.

Two-accommodation zones [169] coexist among these half-grabens which extend transversely across the rift (Figure 10). These are the Galala-Zenima [168] or Gharandal [167] accommodation zone, of broad extension (about 60 km wide) in the north, and the Morgan [168] or Sufr El Dara [170] accommodation zone (20 km wide) in the south. Both zones exhibit a broad range of deformation, including distinct normal, oblique, or strike-slip faults [171], or wide complex zones of normal faulting, trans-tension [172-174] or broad warping [175].

The Gulf of Suez is considered to be an aseismic area during the first half of the last century and this consideration let some researchers (e.g., [176, 177]) to conclude that all the present motion taking place in the Red Sea Rift is transferred into shearing along the DST. Ben-Menahem [178] and Salamon *et al.* [111] studied the seismic activity of the Suez Rift. Fault plane solutions of the m_b 6.1, March 31, 1969 earthquake and other low-magnitude events show that the Gulf of Suez Rift is active which agree with Ben-Menahem and Aboodi [179] results. Considering the tectonic setting, seismicity and earthquake faulting mechanisms, the Gulf of Suez can be divided into three seismogenic sources (Figure 9) as follow.

5.3.1. Southern Gulf of Suez (EG-15) seismogenic source

The EG-15 Southern Gulf of Suez seismogenic source is distinguished by intensive structural deformation. It is characterized by its relatively high seismic activity. The higher seismicity rate at the southern part of the Gulf of Suez is related to the crustal movements among the three surrounding plates: Arabian Plate, African Plate, and Sinai sub-Plate. Six historical events are included in this zone. Those are 28 B.C., 955, 1091, 1195, 1778, and 1839 events. Their intensities range from VI-VII to VIII. The most important event occurred inside this area source is the M_W 6.8, March 31, 1969 Shedwan earthquake [16, 181]. Three foreshocks and 17 aftershocks (m_b 4.5-5.2) located in the Shedwan Island district are related to this big event. However, Maamoun and El-Khashab [182] mentioned that 35 foreshocks, taken place during the last half of March 1969, were preceding the main earthquake. The focal mechanism solutions of the largest two earthquakes (M_W 6.8, March 31, 1969 and M_W 5.5, June 28, 1972 earthquakes) show a normal faulting mechanisms with negligible shear component along the NW-trending fault plane that it is in agreement with the main axis of the Gulf of Suez [183]. This is also consistent

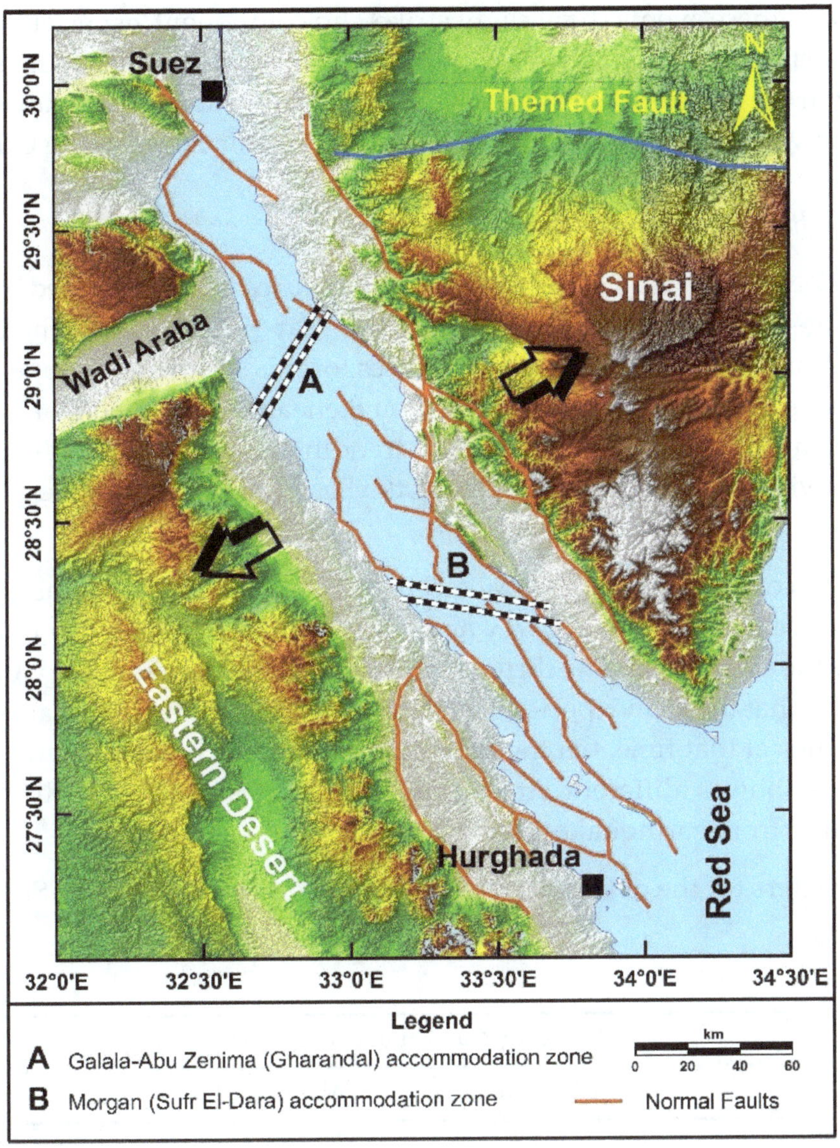

Figure 10. Tectonic setting of the Gulf of Suez. Red lines refer to normal faults (redrawn after Meshref [11]; and Younes and McClay [180]).

with the results obtained using the waveform inversion techniques proposed by Huang and Solomon [184].

5.3.2. Central Gulf of Suez (EG-16) seismogenic source

The seismic activity in the EG-16 Central Gulf of Suez seismogenic source is relatively low when compared with the previous source. Five historical events are included in this source zone: the 1220 B.C., 1425, 1710, 1814, and 1879 earthquakes. Its intensities range from IV to VII. The most important earthquake inside this area was the M_S 6.2 March 6, 1900 event.

Abou Elenean [20] computed some focal mechanism solutions for earthquakes which taken place in the central part of the gulf, showing generally normal faulting, following the main gulf trend. A few of these events show slight strike-slip component, especially for those events closer to the transfer zones of the three gulf dip provinces [11]. This change, from a purely normal faulting in the southern part to a mixed (strike-slip and normal) movement, supports the separation between the southern and middle seismogenic zones in the Gulf of Suez.

5.3.3. Northern Gulf of Suez (EG-17) seismogenic source

Finally, the EG-17 Northern Gulf of Suez seismogenic source is characterized by its low seismic activity. Two large earthquakes occurred before the year 1900. They are the I_{max} VI, 742, and I_{max} V, 1754 earthquakes. Focal mechanism analyses for this seismogenic zone indicate normal faulting mechanism. Fault plane solutions by Abou Elenean [20] showed that the events located at the gulf apex show normal faults, generally trending NW-SE to WNW-ESE, and reflect a good agreement with the surface faults crossing the Eastern Desert from the gulf apex towards Cairo.

Abou Elenean [20] concluded that the focal mechanisms of small to moderate size earthquakes based on the P-wave polarities by Badway and Horváth [185-187], Badawy [188] and Salamon *et al.* [112], show the existence of few thrust faulting mechanisms along the Gulf of Suez trend. The author argues that these unexpected mechanisms could be due to the lack of local stations with clear polarities at that time. On the other hand, borehole breakouts analyses performed by Badawy [188] show a different stress direction, inconsistent with the NE-SW tension direction estimated from earthquake focal mechanisms.

Seismicity parameters for the delineated seismic sources along the Gulf of Suez are displayed in Table 3.

Source Zone	b-value	Yearly Number of Earthquakes		Observed M_{max}
		Above M_W 4.0	Above M_W 5.0	
EG-15	1.06	0.8347	0.0721	M_W 6.8 on 1969/03/31
EG-16	0.80	0.3085	0.0488	M_S 6.2 on 1900/03/06
EG-17	0.86	0.1381	0.0190	M_S 6.6 on 1754/--/--

Table 3. b-value, annual rate of earthquakes, and maximum observed magnitude for the delineated seismic source zones along the Gulf of Suez.

5.4. Seismic sources of the Egyptian Eastern Desert

The Eastern Desert of Egypt, structurally, is a part of the Arabian-Nubian Shield. It lies within the fold and thrust belt of the Pan-African continental margin [189]. It is underlain mainly by the Pre-Cambrian basement of igneous and metamorphic rocks, which constitutes the Nubian Shield that had been formed before the Red Sea opening. It is believed that Nubian Shield basement was stabilized during the Pan-African Orogeny (about 570 Ma ago) [190].

Stern and Hedge [191] divided the Eastern Desert belt into three structural domains (Figure 2): northern, central and southern. These domains are separated by two major faults: i) the first is the Safaga-Qena zone, extending from Safaga to Qena, and ii) the second one is the Marsa Alam-Aswan fault zone. The Eastern Desert is characterized by E-W trending faults in the southern part, which changes to ENE-WSW in the middle one, near to Hurghada city. Further to the north, towards the Cairo-Suez District, the main fault trend becomes in the E-W direction.

However, Youssef [10] classified the main tectonic structures developed in the Eastern Desert into three main groups: i) NW-SE trending normal faults parallel to the Gulf of Suez-Red Sea Rift, ii) NE-SW trending faults parallel to the Gulf of Aqaba, and iii) a set of fault system trending nearly in the E-W direction. In addition, there are many simple and open folds with a NW-SE trend and low plunges.

Deif *et al.* [57] quote that the relationship between the earthquake activity in the Eastern Desert and the causal structures is not fully understood, due to the lack of geological and geophysical studies in this region. Furthermore, no historical earthquakes have been reported in the current seismogenic sources [17, 21]. The following seismogenic sources are identified (Figure 9).

5.4.1. Western Red Sea Coast (EG-18) seismicity source

In addition to the Red Sea seismogenic sources mentioned above, there are some earthquakes located in the region which extends to the west, from the EG-12 Southern Egyptian Red Sea source till the western coast of the Red Sea. This activity may be related to the block adjustment in this region or to some ocean floor spreading. This source is characterized by a low seismic activity. The biggest observed earthquake is the M_L 4.5, May 23, 1990 earthquake.

5.4.2. Southern Eastern Desert (EG-19) seismicity source

This seismicity source exhibit a low seismic activity rate in comparison to the adjacent Red Sea seismic sources. There are no focal mechanism solutions for earthquake events inside this area source. The M_L 4.4, July 15, 1991 earthquake is the biggest recorded event in this zone.

5.4.3. Southern Abu Dabbab (EG-20) seismicity source

Depending on both the changes in the seismicity rate and distribution, another seismicity source (EG-20) has been considered to the north of the previous zone. The same as the previous, there is no focal mechanism solutions in this source. The biggest recorded event is the M_L 4.7, January 21, 1982 earthquake.

5.4.4. Abu Dabbab (EG-21) seismicity source

The Abu Dabbab region is located in the central part of the Eastern Desert of Egypt. The moderate level of seismic activity and extremely tight clustering of low-magnitude earthquakes at Abu Dabbab suggests that the seismicity in this area is not directly related to regional tectonics. One possible explanation is that the activity is related to magmatic intrusions into the Pre-Cambrian crust, but there is no direct evidence to support this hypothesis [161].

The most important event included inside this area source is the M_S 5.3, November 12, 1955 earthquake. This event is felt in the Upper Egypt in Aswan and Qena cities, and as far as Cairo, but no damage was reported. Its focal mechanism solution has normal and strike-slip faulting components produced by a NNW minimum compressive stress and a NE maximum compressive stress. Fault planes strike roughly E-W or N-S to NE-SW. Another important event related to this area is the M_W 5.1, July 2, 1984 earthquake, which is felt strongly in Aswan, Qena and Quseir cities. A large number of foreshocks and a huge sequence of aftershocks are recorded. The focal depth of the whole sequence was less than 12 km.

Seismicity parameters for the delineated seismic sources of the Eastern Desert of Egypt are displayed in Table 4.

Source Zone	b-value	Yearly Number of Earthquakes		Observed M_{max}
		Above M_W 4.0	Above M_W 5.0	
EG-18	1.29	0.1566	0.0080	M_L 4.5 on 1990/05/23
EG-19	1.15	0.1182	0.0085	M_L 3.9 on 1991/07/15
EG-20	1.20	0.0625	0.0039	M_L 4.7 on 1982/01/21
EG-21	0.87	0.3714	0.0496	m_b 6.2 on 1955/11/12

Table 4. b-value, annual rate of earthquakes, and maximum observed magnitude for the delineated seismic source zones of the Eastern Desert.

5.5. Seismic sources along Nasser's Lake, Nile Valley and Cairo-Suez region

5.5.1. Southern Aswan (EG-22) seismogenic source

The geological structural pattern of the Nasser's Lake and Aswan region is characterized by a regional basement rock uplift and regional faulting [192-197]. Faults around the Aswan region, according to their behavior, are grouped into three categories [82]:

i. E-W trending faults (Figure 11), as the Kalabsha and Seiyal Faults, which lay to the west of Nasser's Lake. The Kalabsha Fault is about 185 km long right-lateral strike-slip fault. Its slip rate was estimated to be 0.028 mm/yr., and the Seiyal Fault is considered to be similar to that of the Kalabsha Fault [82].

ii. N-S trending faults (Figure 11), which can be subdivided into two main sets: *The first set* lies to the NW of Nasser's Lake and consists of three faults: the Gebel El-Barqa Fault, the Kurkur Fault and the Khur El-Ramla Fault. The Gebel El-Barqa is a left-lateral strike-slip fault, with a total length of 110 km. The Kurkur Fault is also a left-lateral strike-slip fault, and it is characterized by its low seismic activity if compared with the neighbor faults. The Khur El-Ramla Fault is about 36 km in length, and it has no direct indication of its sense of movement. *The second set* of faults are lying to the SW of Nasser's Lake, and consists mainly of two faults: the Abu Dirwa and the Ghazala Faults. Abu Dirwa Fault is a 20 km long left-lateral strike-slip fault and it has a very low degree of seismic activity. In addition, for the Gazelle Fault, the analysis of its geomorphic expression shows no active features, and that there is no ground cracks observed along the fault trace. Likely this fault is inactive [192]. The fault planes of this system are nearly vertical (80-85°).

iii. The third one is a fault system trending NNE-SSW (Figure 11) and lies to the east of Nasser's Lake. The Dabud Fault, which represents the main fault of this group, is about 36 km length. Geological evidences indicate reverse-slip, opposed to the tectonic setting of the area.

In addition to the previous fault systems, Deif [57] has mentioned three faults located at the High Dam area. They are the Powerhouse, the Spillway and the Channel Faults. Deif [57] provided that the evidence of the occurrence of these faults is hidden below the Aswan High Dam and Nasser's Lake. These three faults show no evidence of being active in the Quaternary, and are considered as inactive with no significant hazard to the Aswan region [82].

No historical earthquakes were reported by Ambraseys *et al.* [18] inside this area source. However, two historical events (epicentral intensity VII) were reported by Maamoun *et al.* [16] to be located at the same place of the M_W 5.8, November 14, 1981 earthquake. These two events occurred in 1210 B.C. and in 1854.

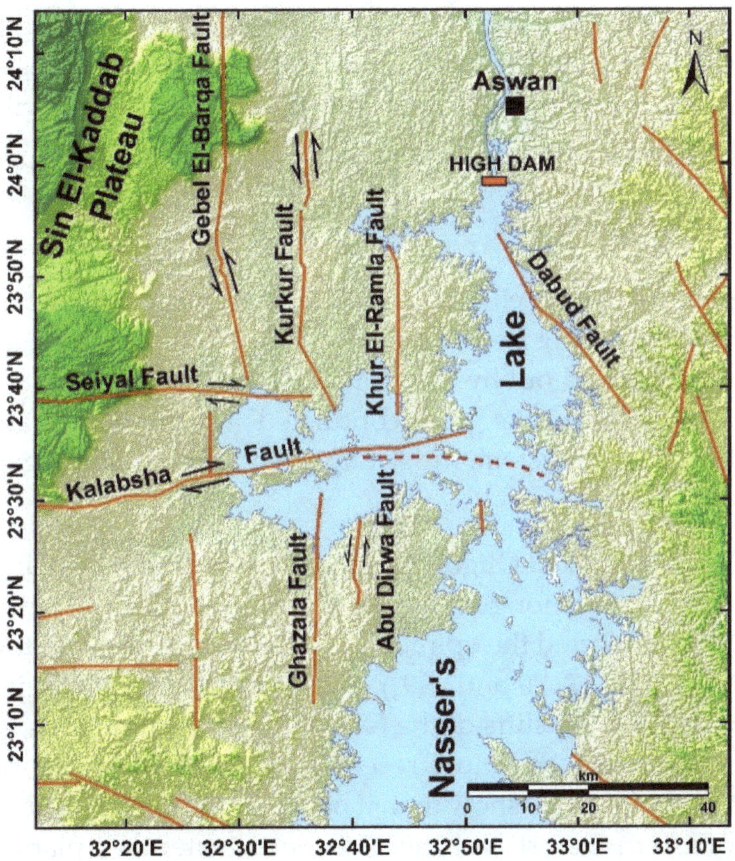

Figure 11. Geological and tectonic features around Nasser's Lake (redrawn after Woodward-Clyde Consultants [82]).

Woodward-Clyde Consultants [82] evaluated the fault system in the Kalabsha area and reported that the Western Desert Fault System consists of a set of E-W faults that exhibit dextral-slip displacement, and a set of N-S faults that exhibit sinistral-slip displacement. The E-W faults are longer, and have greater degree of activity in the Quaternary, having larger total slip rates (about 0.03 mm/yr.) than the N-S faults (0.01–0.02 mm/yr.).

Many seismic hazard studies have been carried out in the Aswan area and its surroundings due to its importance and neighborhood to the High Dam (e.g., [51, 57 and 58]). Three alternative seismotectonic models for Aswan area have been considered in these studies. The first model consider the Aswan Area as one seismotectonic model, while in the second one is subdivided into six seismotectonic provinces. The third model is mainly depending on the fault seismic sources. The latter one is based mainly on the well-known defined active faults and its associated seismic activities.

However, this work, the Aswan region and its surroundings is considered as one source zone (Figure 9). The main earthquake that took place inside this area was the M_W 5.8, November 14, 1981 event. This earthquake occurred in the Nubian Desert of Aswan. It is of great significance because of its possible association with Nasser's Lake. Its effects were strongly felt up to Assiut city (440 km to the north from Kalabsha Fault), as well as to Khartoum city (870 km to the south). Several cracks on the western bank of the Nasser's Lake, and several rock-falls and minor cracks on the eastern bank, are reported. The largest of these cracks is about one meter in width and 20 km in length. This earthquake was preceded by three foreshocks and followed by a large number of aftershocks. The focal depth of this earthquake is estimated to be 25 km. The composite fault plane solution of this event indicates a nearly pure strike-slip faulting with a normal-fault component [49, 195].

5.5.2. Luxor- Southern Beni Suef (EG-23) seismogenic source

Several geophysical studies have been carried out by many authors using different approaches in individual localities lying along the Nile Valley. The most interesting geological studies in the Nile Valley are those carried out by Said [69-71], Issawi [78], Philobbos *et al.* [86], and El-Younsy *et al.* [196]. All these works were conducted independently and aimed to obtain information about the drainage system, the stratigraphy and structural geology in this part of Egypt.

The Nile Valley is a large elongated Oligo-Miocene rift, trending N-S as an echo of the Red Sea rifting. There is no agreement among scientists, till now, about the origin of the Nile Valley. Some authors [197, 198] supported the opinion of the erosional origin of the Nile Valley, while many others (e.g., [11, 12, 13, 17, 70 and 199]) consider the tectonic origin. This is supported by the fault scarps bordering the cliffs of the Nile Valley, the numerous faults recognized on its sides [70, 71 and 199] and the most recent focal mechanism solutions. Furthermore, geological studies of the Nile Valley show that, it occupies the marginal area between two main tectonic blocks (the Eastern Desert and the Neogene-Quaternary platform), which in turn behaves as a barrier that prevents the further extension of the East African Orogenic Belt activity to the west [71].

From the structural point of view, the faults and joints are the most deformational features observed at the cliffs bordering the Nile stream [69 and 70]. These faults have different directions (Figure 2). The most abundant present the NW-SE and NNW-SSE trends, while others (less abundant) exhibit the WNW-ESE, ENE-WSW and NE-SW directions. Most of the major valleys, at the east of the Nile River, are generated and controlled in a more or less degree by these faults.

To the north of Aswan area, in the region between Luxor and Southern Beni Suef, along the Nile River, there is a low seismicity level, which coincides with the main trend of the Nile River. This active area has been considered as a separate seismogenic source. Several historical earthquakes are reported to occur along the Nile River in this area source that may be due to the high population density along the Nile River in the ancient times. These earthquakes are the 600 B.C., 27 B.C., 857, 967, 997, 1264, 1299, 1694, 1778, and 1850 events. Their intensities range from V to VIII. Focal mechanism solutions exhibit reverse faulting mechanism to the west of the Nile River, in the area between Luxor and Assiut. However, normal faulting mechanism with strike-slip component appears to the north of Assiut till Beni Suef city.

5.5.3. Beni Suef – Cairo – Suez District (EG-24) seismogenic source

To the north of the previous zone and to the west of the Gulf of Suez, there is a moderate seismic activity between Beni Suef and Cairo, on the River Nile, till Suez, on the apex of the Gulf of Suez (Figure 9). Three fault trends are affecting the Cairo-Suez district: the first one is trending E-W, which aligned by latitude 30°N, and it is very dominant, while the other two (ENE and NW) are spatially more abundant [200]. The faults are predominantly normal, and have produced a series of fault blocks with a large strike-slip component [200].

Field observations, satellite images, aerial photographs and seismic profiles confirm that the region between Cairo and Suez is active from a tectonic point of view. Seismic activity are noticed along this belt at Wadi Hagul and Abu Hammad. However, the earthquake distribution in this area is very scattered, and cannot be attributed to a specific known fault. This disperse seismicity yields a difficulty in delineating seismic zones. It is assumed that the seismic potential is uniform throughout the zone, although this is not entirely clear.

Sixty one historical earthquakes are related to this area source. The most important among them are the 935, 1111, 1259, 1262, 1303, and 1588 events. Moreover, the most important instrumental earthquake taken place in this source is the M_W 5.8, October 12, 1992 event. Its epicenter was located about 40 km south of Cairo, in Dahshour. It caused a disproportional damage (estimated at more than L.E. 500 million) and the loss of many lives. The shock was strongly felt, and caused sporadic damage and life loss in the Nile Delta, around Zagazig. Damage was extended to reach Fayoum, Beni Suef and Minia cities. The mostly affected area was Cairo, especially its old sections, Bulaq and the southern region, along the western bank of the Nile to Gerza (Jirza) and El-Rouda. In all, 350 buildings collapsed completely and 9000 were irreparably damaged, killing 545 persons and injuring 6512. Most causalities in Cairo were victims of the horrible stampedes of students rushing out from schools. Approximately, 350 schools and 216 mosques were destroyed and there was about 50000 homeless.

Abdel Tawab *et al.* [201] studied the surface tectonic features of the area around Dahshour and Kom El-Hawa, and found a major N55°E trending normal fault at Kom El-Hawa (800 m length of surface trace with a vertical displacement of 40 cm) and a major E-W trending open fracture at Dahshour area (1200 m in length). Maamoun *et al.* [202] concluded that, most of the surface lineaments recorded after the occurrence of the main shocks are trending E-W to NW-SE. Abou Elenean [19] studied the focal mechanism solutions for some earthquakes in Dahshour area, and found normal faulting with a large strike-slip component. The first nodal plane is trending nearly E-W, showing coincidence with the surface lineaments that appeared directly after the occurrence of the M_S 5.9, 1992 earthquake.

In addition to the previous earthquake, there were three important earthquakes located inside this source zone. One of them located to the southwest of Suez, is the m_b 4.9 (M_S 4.8), March 29, 1984 Wadi Hagul earthquake. It was strongly felt in Suez, Ismailia and Cairo. A large number of aftershocks were recorded by nearby temporary stations. The second earthquake was located Northeast Cairo; it is the m_b 4.8, April 29, 1974 Abu Hammad earthquake. It was strongly felt in Lower Egypt (Nile Delta) and Southwest Israel. The last earthquake was the m_b 5.0, January 2, 1987 Ismailia event.

Mousa [203] and Hassib [204] computed two nodal planes trending ENE-WSW and NNW-SSE, with left-lateral strike-slip motion along the second plane, for the Abu Hammad event. They computed the same strike-slip with reverse component for the Wadi-Hagul earthquake. In addition, the mechanism of the Ismailia earthquake shows also strike-slip movement with two nodal planes trending N68° E and S24°E, with steep dip angles (each of them is 80°) [205].

The seismicity parameters for the delineated seismic sources along the Nile River are displayed in Table 5.

Source Zone	b-value	Yearly Number of Earthquakes		Observed M_{max}
		Above M_W 4.0	Above M_W 5.0	
EG-22	0.79	0.5860	0.0946	M_W 5.8 on 1981/11/14
EG-23	0.73	0.2948	0.0549	I_{max} VIII on 0857/04/--
EG-24	0.99	0.5964	0.0608	I_{max} IX-X on 1262/--/--

Table 5. b-value, annual rate of earthquakes, and maximum observed magnitude for the delineated seismic source zones along the Nile River.

5.6. Seismic sources along the Mediterranean Coastal Line

The Mediterranean Coastal area is characterized by small to moderate seismicity. This area is located at the southeastern part of the Mediterranean Sea. It separates between the high seismic activities along the Gulf of Aqaba-Dead Sea Transform Fault and the seismicity of the Mediterranean Sea (Hellenic and Cyprian Arcs). Moreover, it separates the Southern Cyprus seismic activity from the Northern Egypt activity. Hence, this area has been divided into three seismic sources (Figure 12), based mainly on the available focal mechanism data and the seismic activity.

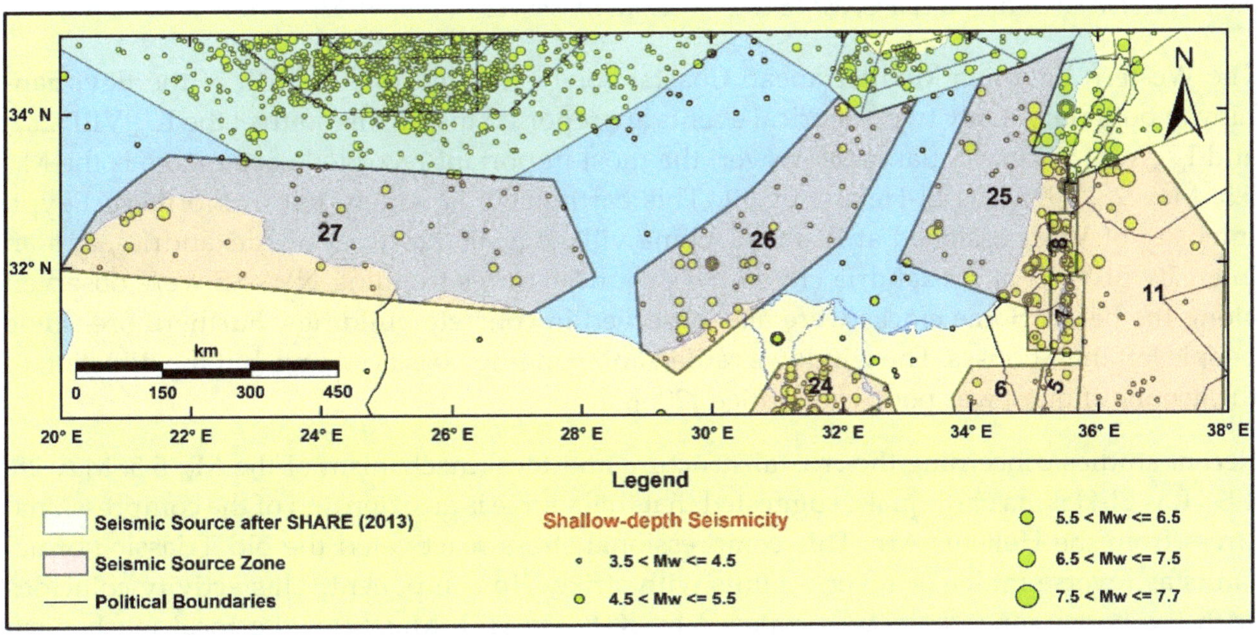

Figure 12. Shallow-depth seismicity (h ≤ 35 km) and seismic source zones delineated for the Mediterranean coastal line.

5.6.1. Eastern Mediterranean Coast (EG-25) seismicity source

This area source is parallel to the eastern coastal line of the Mediterranean Sea. It is located to the west of the previous quoted sources EG-07, EG-08, and EG-09, and to the southeast of Cyprus. It includes all the seismicity located to the west of the DST, and those earthquakes are not related with the Cyprian Arc. 29 historical events are included inside this area source. The most important among them are the 590 B.C., 525 B.C., 12 B.C., 306, 332, 551, 1269, and 1546 earthquakes.

5.6.2. Northern Delta (EG-26) seismicity source

This source is located to the northwest of the Nile Delta region. It extends from Alexandria towards the Mediterranean Sea in NE direction. 23 historical events are included inside this large area source, among them the 796, 951, 955, 956, 1303, 1341, and 1375 earthquakes. Moreover, the m_b 6.5 (M_S 6.8), September 12, 1955 Alexandria earthquake, represents the most important recorded event inside this source. This earthquake was felt in the entire Eastern Mediterranean Basin. In Egypt, it was strongly felt, and led to the loss of 22 lives and damage in the Nile Delta, between Alexandria and Cairo [17]. The destruction of more than 300 buildings of old brick construction was reported in Rosetta, Idku, Damanhour, Mohmoudya and Abu-Hommos. A maximum intensity of VII was assigned to a limited area in Behira province, where 5 people killed and 41 were injured.

Mostly of the focal mechanism data inside this area source reflects reverse faulting mechanism with, sometimes, strike-slip component, except one event, showing a strike-slip motion with a notable normal component (the M_W 4.5, April 9, 1987 event).

5.6.3. *Western Mediterranean Coast (EG-27) seismicity source*

The Western Egyptian Mediterranean Coastal zone is located to the north of the Egyptian-Libyan boundary. Only two historical events are reported inside this source: the I_{max} VIII, 262, and I_{max} VI, 1537 earthquakes. However, the most important recorded earthquake is the M_W 5.5, May 28, 1998 Ras El-Hekma event. This earthquake is widely felt in Northern Egypt. Intensity of VII is assigned at Ras El- Hekma village (~300 km west of Alexandria), and an intensity of V–VI at Alexandria city [206]. Ground fissures trending NW–SE were observed along the beach. Some cracks were also observed in concrete buildings. Furthermore, some people left their houses. The windows rattled and hanging objects swung, but the direction of the ground motion was poorly identified [206].

Recent studies concerning the crustal structure and focal mechanism of the M_W 5.5, May 28, 1998 Ras El-Hekma earthquake suggested that this source is an extension of the compressional stress from the Hellenic Arc. This compressional stress reactivated the old Triassic normal faults as reverse faults, or reverse faults with strike-slip component. This activity coincides with the hinge zone geometry proposed by Kebeasy [17]. Mostly of the focal mechanism analyses data indicate reverse faulting with some strike-slip component.

Seismicity parameters for the delineated seismic sources along the Mediterranean coastal line are displayed in Table 6.

Source Zone	b-value	Yearly Number of Earthquakes		Observed M_{max}
		Above M_W 4.0	Above M_W 5.0	
EG-25	0.97	0.5204	0.0554	I_{max} X on 1546/01/14
EG-26	0.94	0.5975	0.0688	m_b 6.5 on 1955/09/12
EG-27	0.60	0.7134	0.1793	I_{max} VIII on 0262/--/--

Table 6. b-value, annual rate of earthquakes, and maximum observed magnitude for the delineated seismic source zones along the Mediterranean coastal line.

5.7. Seismic sources of the Western Desert

5.7.1. *El-Gilf El-Kebeir (EG-28) Seismogenic Source*

Issawi [28] studied the geology of El-Gilf El-Kebeir region, and concluded that the area is affected by three main faults (Figure 2). The first one is the Gilf Fault, which strikes N-S for a distance of 150 km inside Egypt. Its extension in Sudan is unknown. Its northward extension is not traced. He interpreted this fault as a normal, gravity, strike and hinge type of structure. The second one is Kemal Fault, which limits the northwestern side of the Gilf Plateau. It is normal, strike fault which trends NW-SE. The Kemal Fault intersects the Gilf Fault at its northern end. The third one is the Tarfawi Fault, which has the same trend similar to the Gilf Fault. Its length, in Egypt, is 220 km but it extends in Sudan. He interpreted this fault as a normal, gravity and hinge fault.

The only recorded earthquake in this area source is the m_b 5.3 (M_L 5.7), December 9, 1978 El-Gilf El-Kebeir earthquake. It had a reverse faulting mechanism. Riad and Hosney [207] studied its focal mechanism and concluded that, a shear direction did exist in the basement rocks of the southern part of the Western Desert and has been explained as due to compressional stress resulting from the spreading of the Red Sea. Their fault planes solution shows that the P-axis is almost perpendicular to the Red Sea spreading axis. They concluded that the Gilf Plateau is probably divided into two parts by a fault striking nearly E-W. Some authors [e.g., 208 and 209] pointed out that this activity is linked to the pre-existing weak zones, while, Abou Elenean [19] linked such an intraplate activity to the intersection of more than one local fault.

In the current work, the Gilf El-Kebeir (EG-28) seismogenic source covers the seismic activity in this area, as well as the above mentioned faults.

6. Eastern Mediterranean region seismic sources

The Mediterranean region is characterized by a very complex tectonics that can be generally described in the frame of the collision between the Eurasian and African Plates [183, 210-219]. It can be divided into western, central, and eastern basins.

The Eastern Mediterranean region, which defines the region lying between the Caspian Sea and the Adriatic Sea through Caucasus, Anatolia, Aegean Sea and Greece, is one of the world's most seismically active regions. Recent tectonics of the Eastern Mediterranean region has been studied intensely in the last four decades. The Eastern Mediterranean region is known to be seismically active over a period of more than 2000 years based on historical and instrumental records. The tectonic and seismotectonic studies reflect a highly complicated tectonic setting.

It is characterized by two main seismic regions: the Hellenic and Cyprian Arcs (Figure 13). The Cyprian Arc has a similar geometry to the Hellenic Arc and the two are often compared (e.g., [220]). However, the observed seismic activity and the well-known plate movement in the Eastern Mediterranean area, suggest that the previously mentioned arcs are affected by a very distinct tectonic activity. The convergence across the first one (Hellenic Arc) is 20–40 mm/yr. (two to three times faster across the Cyprian Arc). Thus, this biggest displacement level yields in higher seismicity rate at much deeper levels (up to 300 km) [220].

The Cyprian Arc represents a tectonic plate margin separating the Anatolian sub-Plate (to the north) from the Nubian and Sinai sub-Plates (to the south) (Figure 13). It is connected from the west by the Hellenic Arc, and from the east by the Dead Sea and the East Anatolian Faults. In addition, it extends from the Gulf of Antalia, to the west to the Gulf of Iskenderun, to the east. On the other hand, the Hellenic Arc is considered to be the most active seismic region in Europe. It is represents the convergent plate boundary between the African Plate and the Eurasian Plate (Aegean sub-Plate) in the Mediterranean area (Figure 13).

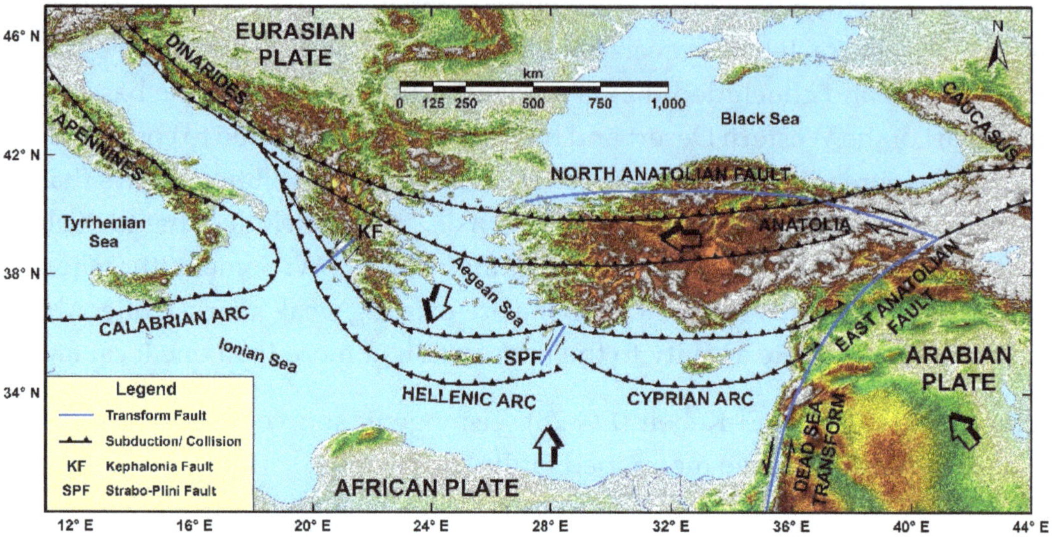

Figure 13. Tectonic map of the Eastern Mediterranean region (after Ziegler [221]; Meulenkamp *et al.* [222]; and Dewey *et al.* [223]).

6.1. SHARE shallow-depth seismic sources (h ≤ 20 km)

The Seismic Hazard Harmonization in Europe (SHARE) project [224], since the year 2009 till 2013, worked in establishing an appropriate seismic hazard model for Europe and Turkey. This project delivered a seismic hazard reference model for the current use of the European building design and seismic regulations, Eurocode 8 (EC8), that came into force in 2010.

The EU-FP7 European Commission Project (SHARE), aiming at providing an updated state-of-the art time-independent seismic hazard model, envisioned to serve as a reference model for the revision of the EC8 building code. SHARE, in addition, contributes its results to the Global Earthquake Model (GEM, www.globalquakemodel.org), a public/private partnership initiated and approved by the Global Science Forum of the OECD- GSF, aiming to provide a uniform hazard and risk model around the globe.

The Euro-Mediterranean area is complex from a seismotectonic point of view. The plate boundary between Africa and Europe runs roughly west to east from the Mid-Atlantic Ridge to Eastern Turkey with different mechanisms including continental collision, subduction, and transcurrent movement. Moving away from the plate boundary, the stable continental region is also locally rather active.

SHARE inherits knowledge from national, regional and site-specific PSHAs, assessed new data, assembled the data in a homogeneous fashion, and built comprehensive hazard relevant databases. In the frame of this project, the establishment of a seismic source model for Europe and the surrounding areas was considered. This model is built upon the available local and regional models as well as newly defined source zones. It has been developed during eight separate workshops by the SHARE consortium. Almost 80 experts from 28 countries from the informed European-Mediterranean seismological community have participated in building the zonation model.

The principle for seismic source zones is that they represent enclosed areas within which, a uniform seismicity distribution and maximum magnitude is expected. Background sources have been avoided in the sense that all areas have been covered by seismic sources, i.e., even very low seismicity areas are covered with areal source zones. The principles along which seismic source zones in the current model have been constructed are based on information from geological structures on different scales, tectonics and seismicity.

Seismicity also follows these structures well, e.g., as can be seen along the North Anatolian Fault, the Gulf of Corinth and the Hellenic Arc. The use of fault source information has also been done in the delineation of the source zones, especially in the case of the foundation of the sources for Balkans, Greece and Turkey, Italy and Portugal. b-value, annual activity rates, and maximum expected magnitude were computed using different approaches and methods and included in the SHARE project database (www.share-eu.org) [225].

In the current work, 53 shallow-depth seismic sources ($h \leq 20$ km) from the SHARE source model (Figure 5), were considered to the north of Egypt, till latitude 38° N, and covering the Greece and Turkey regions. Some of the events located at this region were felt and caused few damages in the northern part of Egypt (e.g., the I_{max} VIII, August 8, 1303 offshore Mediterranean earthquake, the I_{max} VI, February 13, 1756 and the M_S 7.4, June 26, 1926 Hellenic Arc earthquakes, the M_W 6.8, October 9, 1996 Cyprus earthquake, and the M_S 6.4, October 12, 2013 Crete earthquake). Thus, these source zones have a certain contribution to the seismic hazard in Northern Egypt.

The model in the Greek and Cyprian area build to a large extent upon the previous works of Papiouannou and Papazachos [64] and Papiouannou [226]. The Turkish model [227, 228] is provided as a cooperation between the EMME project and SHARE. The Turkish model is further a hybrid model, in the sense that the area sources have been delineated with respect to the integrated fault line sources from the main faults, like the North and East Anatolian Faults.

6.2. Intermediate-depth seismic sources ($20 \leq h \leq 100$ km)

Intermediate focal depth earthquakes occur in the Eastern Mediterranean region (Southern Greece and Turkey) define a Benioff zone of stair shape which dips from the convex side of the Cyprian and the Hellenic Arcs to its concave side (from the Eastern Mediterranean to the Greek and Turkish lands) [229-231]. Some of these earthquakes are moderate to large earthquakes, and constitute a seismic threat for the whole Mediterranean area, including Northern Egypt. Since, because of their magnitudes and focal depths, these earthquakes produce seismic waves of large amplitude and period which travel large distances with low attenuation [232]. Therefore, these earthquakes can contribute to the seismic hazard of Northern Egypt.

In this work, intermediate-depth sources for earthquakes having focal depths ranging from 20 km to 100 km have been delineated. Below this depth (100 km) and considering the large distance from Egypt, deep events have no contribution to the seismic hazard. Thus, 7 intermediate-depth source zones have been considered in the Hellenic and Cyprian subduction zones to cover the intermediate-depth seismicity ($20 \leq h \leq 100$ km) (Figure 14). The zoning was based on the seismicity distribution and the tectonic setting of the region. Seismicity parameters for these intermediate-depth seismic sources are displayed in Table 7.

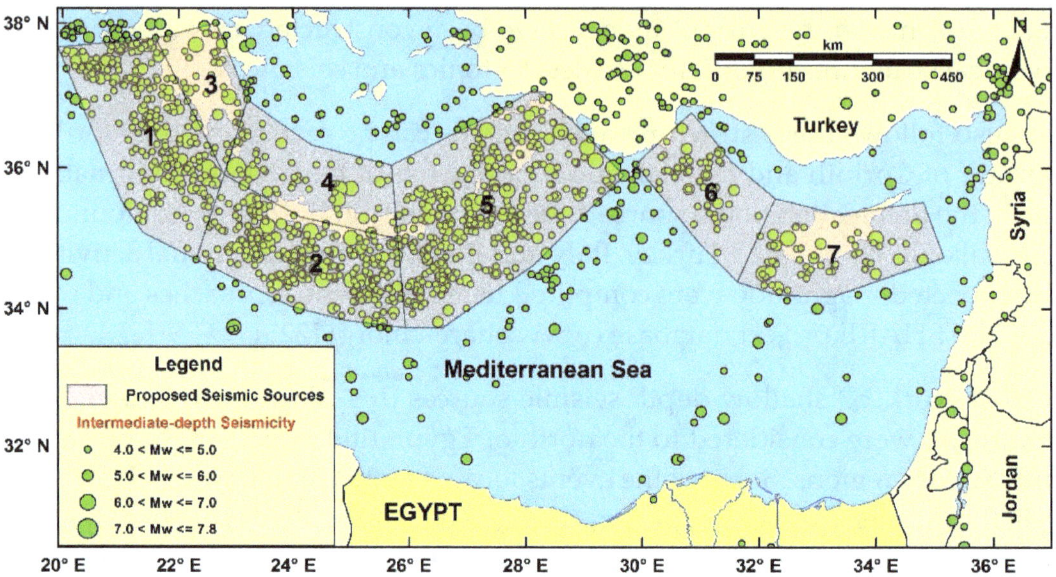

Figure 14. Intermediate-depth seismicity (20 ≤ h ≤ 100 km) and delineated intermediate-depth seismogenic source zones for the Eastern Mediterranean region.

Source Zone	b-value	Yearly Number of Earthquakes		Observed M_{max}
		Above M_W 4.0	Above M_W 5.0	
MD-01	0.88	12.2091	1.6212	M_W 6.9 on 2008/02/14
MD-02	0.97	14.1395	1.5283	m_b 6.4 on 2013/10/12
MD-03	0.78	1.9892	0.3324	m_b 6.9 on 1962/08/28
MD-04	0.83	2.8908	0.4316	m_b 7.2 on 1903/08/11
MD-05	0.88	14.8782	1.9434	m_b 7.1 on 1984/02/09
MD-06	0.93	2.2986	0.2727	M_S 5.8 on 1984/04/30
MD-07	0.82	2.9083	0.4409	m_b 6.5 on 1941/01/20

Table 7. b-value, annual rate of earthquakes, and maximum observed magnitude for the delineated seismic source zones of the Eastern Mediterranean region.

7. Conclusion

To reach a more realistic seismic hazard quantification in Egypt, it is necessary to recognize the seismic source zones, including the seismic activity that can affect different regions all over the country. In the current work, a new seismic source model for Egypt and its surroundings is proposed, using all available geological, geophysical, tectonic and earthquake data, aimed at carrying out seismic hazard studies.

This work presents a detailed review on major tectonic features and the correlation of seismicity with them, to demarcate seismic sources in Egypt and neighborhood. The Gulf of Aqaba-Dead Sea Transform, the Red Sea-Gulf of Suez Rift, and the Cyprian and Hellenic Arcs are the three most active seismotectonic belts in the region, which have produced several large earthquakes in the recent past. On the basis of a comprehensive and critical analysis of the seismotectonic characteristics, different seismic sources are defined to model the seismicity for the assessment of seismic hazard in Egypt.

Focal mechanism solutions data, active faults data, as well as an updated earthquake catalogue for the period 2200 B.C.–2013 are taken into account. Potential seismic sources are modeled as area sources, in which configuration of each seismic source is controlled, mainly, by the fault extension and seismicity distribution.

The proposed seismic source model consists of 28 shallow-depth seismic zones ($h \leq 35$ km) for the Egyptian territory and its surroundings, specified on the basis of mainly seismotectonic and seismicity criteria. In addition, the authors have considered 53 shallow-depth seismic sources ($h \leq 20$ km) for the Eastern Mediterranean region after SHARE (2013). Furthermore, the current model involves 7 delineated intermediate-depth seismic sources ($20 \leq h \leq 100$ km) covering the intermediate-depth seismicity in the Eastern Mediterranean region.

Seismicity parameters (b-value and activity rates) of the Gutenberg–Richter magnitude–frequency relationship have been estimated for each one of the seismic sources. In addition, the maximum observed magnitude for each seismic source zone was reported from the sources sub-catalogues. The coordinates of these seismic source zones and the estimated seismicity parameters can be directly inserted into PSHA after the estimation of the maximum expected magnitude for each source. The computation of seismic hazard for Egypt using these data will form the subject matter of a future paper.

Acknowledgements

The first author wants to thank the Egyptian Government for funding him in the Joint-Supervision Mission program at the University of Jaén, Spain. This research was supported by the Aswan Regional Earthquake Research Centre and the Spanish Seismic Hazard and Active Tectonics research group.

Author details

R. Sawires[1,4], J.A. Peláez[2], R.E. Fat-Helbary[3], H.A. Ibrahim[1] and M.T. García Hernández[2]

1 Department of Geology, Faculty of Science, Assiut University, Assiut, Egypt

2 Department of Physics, University of Jaén, Jaén, Spain

3 Aswan Seismological Center, Aswan, Egypt

4 Department of Physics, University of Jaén, Jaén, Spain

References

[1] Cornell CA. Engineering Seismic Risk Analysis. Bulletin of Seismological Society of America 1968; 18: 1583-1606.

[2] Reiter L. Earthquake Hazard Analysis. Columbia University Press, 1990.

[3] McGuire RK. FORTRAN Computer Programs for Seismic Risk Analysis. United States Geological Survey. Open-File Report No. 76-67, 1976.

[4] Kramer SL. Geotechnical Earthquake Engineering. Prentice-Hall Editor. Upper Saddle River, New Jersey, 07458; 1996.

[5] Abrahamson N. Seismic Hazard Assessment: Problems with Current Practice and Future Development. First European Conference on Earthquake Engineering and Seismology, 3-8 September 2006, Geneva, Switzerland.

[6] McGuire RK. Computer Program for Seismic Risk Analysis Using Faults as Earthquake Sources (FRISK) 1978. US. Department of Interior Geological Survey. Open-File Report 78-1007, 71 pp., Denever, Colorado.

[7] Bender BK. and Perkins DM. SEISRISK III: A Computer Program for Seismic Hazard Estimation. USGS Bulletin 1987: 1772.

[8] Ordaz M., Faccioli E., Martinelli F., Aguilar A., Arboleda J., Meletti C., and D'Amico V. CRISIS 2014. Institute of Engineering, UNAM, Mexico City, Mexico.

[9] Woo G. NPRISK Seismic Hazard Computation Algorithm Based on Cornell-McGuire Principles. Code developed at NORSAR; 1994.

[10] Youssef MI. Structural Pattern of Egypt and its Interpretation. The American Association of Petrolum Geologists Bulletin 1968; 53: 601-614.

[11] Meshref W. Tectonic Framework. In: Said R. (ed.). The Geology of Egypt. A.A. Balkema, Rotterdam, Netherlands; 1990. p113-155.

[12] Sieberg A. Handbuch der Geophysik. Band IV, Erdbeben-geographie. Borntraeger, Berlin, 1932a. p527-1005.

[13] Sieberg A. Erdbeben und Bruchschollenbau in Östlichen Mittelmeergebiet. Denkschriften der Medizinisch-Naturwissenschaftlichen Gesellschaft zu Jena 18, No. 2; 1932b.

[14] Ismail A. Near and Local Earthquakes at Helwan from 1903-1950. Helwan Observatory Bulletin No. 49, 1960.

[15] Gergawi A. and Khashab A. Seismicity of U.A.R. Helwan Observatory Bulletin No. 76, 1968.

[16] Maamoun M., Allam A. and Megahed A. Seismicity of Egypt. Bulletin of Helwan Institute of Astronomy and Geophysics, 109-160, 1984.

[17] Kebeasy RM. Seismicity. In: Said R. (ed.) The Geology of Egypt. A.A. Balkerma, Rotterdam, Netherlands; 1990. p51-59.

[18] Ambraseys NN., Melville CP., and Adams RD. The seismicity of Egypt, Arabia and Red Sea. Cambridge University Press; 1994.

[19] Abou Elenean K. Seismotectonics of Egypt in relation to the Mediterranean and Red Sea tectonics. PhD. thesis. Ain Shams University, Egypt; 1997.

[20] Abou Elenean K. Focal Mechanism of Small and Moderate Size Earthquakes Recorded by the Egyptian National Seismic Network (ENSN), Egypt. NRIAG Journal of Geophysics 2007; 6 (1) 119-153.

[21] Badawy A. Historical Seismicity of Egypt. Acta Geodaetica et Geophysica Hungarica 1999; 34 (1-2) 119-135.

[22] Badawy A. Seismicity of Egypt. Seismological Research Letters 2005; 76 (2) 149-160.

[23] Abou Elenean KM., Mohamed AME., and Hussein HM. Source Parameters and Ground Motion of the Suez-Cairo Shear Zone Earthquakes, Eastern Desert, Egypt. Natural Hazards 2010; 52: 431-451.

[24] Ziegler MA. Late Permian to Holocene Paleofacies Evolution of the Arabian Plate and its Hydrocarbon Occurrences. GeoArabia 2001; 6 (3) 445-504.

[25] Pollastro RM. Total Petroleum Systems of the Paleozoic and Jurassic, Greater Ghawar Uplift and Adjoining Provinces of Central Saudi Arabia and Northern Arabian-Persian Gulf. US. Geological Survey Bulletin 2202-H; 2003.

[26] Egyptian Geological Survey and Mining Authority "EGSMA". Geologic Map of Egypt 1:2000000; 1981.

[27] Riad S. Shaer Zones in North Egypt Interpreted from Gravity Data. Geophsyics 1977; 24 (6) 1207-1214.

[28] Issawi B. New Findings on the Geology of Uweinat, Gilf Kebir, Western Desert, Egypt. Annals of the Geological Survey of Egypt 1978; 8: 275-293.

[29] Maamoun M. and Ibrahim EM. Tectonic Activity in Egypt as Indicated by Earthquake. Helwan Institute of Astronomy and Geophysics. No. 170, 1978.

[30] Albert RNH. Seismicity and Earthquake Hazard at the Proposed Site for a Nuclear Power Plant in the El-Dabaa Area, North Western Desert, Egypt. Acta Geophysica Polonica 1986; 34 (3) 263-281.

[31] Albert RNH. Seismicity and Earthquake Hazard at the Proposed Site for a Nuclear Power Plant in the Anshas Area, Nile Delta, Egypt. Acta Geophysica Polonica 1987; 35 (4) 343-363.

[32] Kebeasy RM., Maamoun M., and Albert RNH. Earthquake Activity and Earthquake Risk around the Alexandria Area in Egypt. Acta Geophysica Polonica 1981; 29 (1) 37-48.

[33] Kebeasy R. Seismicity of Egypt. Personal Communication; 1984.

[34] Marzouk I. Study of the Crustal Structure of Egypt Deduced from Deep Seismic and Gravity Data. PhD. thesis. Institute of Geophysics, University of Hamburg; 1988.

[35] Fat-Helbary RE. Investigation and Assessment of Seismic Hazard in Egypt. Unpublished Report Submitted to MAPFRE, Spain; 1999.

[36] Fat-Helbary RE. Assessment of Seismic Hazard and Risk in Aswan Area, Egypt. PhD. thesis. Tokyo University, Japan; 1994.

[37] Fat-Helbary RE. Probablistic Analysis of Potential Ground Motion Levels at the Principal Cities in Upper Egypt. Journal of Applied Geophysics 2003; 2: 279-286.

[38] Reborto P., Paolo L., and Dimitris D. Seismotectonic Regionalization of the Red Sea Area and its Application to Seismic Risk Analysis. Natural Hazard 1992; 5: 233-247.

[39] Mohammed A. Seismic Microzoning Study and its Applications in Egypt. PhD. thesis. Ain Shams University, Egypt; 1993.

[40] EL-Hadidy S. Crustal Structure and its Related Causative Tectonics in Northern Egypt using Geophysical Data. PhD. thesis. Ain Shams University, Egypt, 1995.

[41] Fat-Helbary RE., and Ohta Y. Assessment of Seismic Hazard in Aswan Area, Egypt. 11th World Conference on Earthquake Engineering. Paper No. 136 Published by Elsevier Science Ltd 1996.

[42] El-Sayed A. and Wahlström R. Distribution of the Energy Release, b-values and Seismic Hazard in Egypt. Natural Hazards1996; 13: 133-150.

[43] Abou Elenean K. Seismotectonics Studies of El-Dabaa and its surroundings. Unpublished Report 2010. NRIAG, Egypt.

[44] Badawy A. Earthquake Hazard Analysis in Northern Egypt. Acta Geodaetica et Geophysica Hungarica 1998; 33 (2-4) 341-357.

[45] Deif A. Seismic Hazard Assessment in and around Egypt in Relation to Plate Tectonics. PhD. thesis. Ain Shams University, Egypt, 1998.

[46] Riad S., Ghalib M., El-Difrawy MA., Gamal M. Probabilistic Seismic Hazard Assessment in Egypt. Annals of the Geological Survey of Egypt 2000; 23: 851-881.

[47] Abou ELenean K. and Deif A. Seismic Zoning of Egypt. Unpublished Work 2001. NRIAG, Egypt.

[48] El-Sayed A., Vaccari V. and Panza GF. Deterministic Seismic Hazard in Egypt. Geophysical Journal International 2001; 144: 555-567.

[49] Fat-Helbary RE., and Tealeb AA. A Study of Seismicity and Earthqauake Hazard at the Proposed Kalabsha Dam Site, Aswan, Egypt. Natural Hazards 2002; 25: 117-133.

[50] El-Amin EM. Study of Seismic Activity and its Hazard in Southern Egypt. MSc. thesis. Assiut University, Egypt; 2004

[51] El-Amin EM. Study of Seismic Hazard Analysis Using Fault Parameter Solutions in Aswan Region, Upper Egypt. PhD. thesis. Assiut University, Egypt; 2011.

[52] El-Hefnawy M., Deif A., El-Hemamy ST., and Gomaa NM. Probablistic Assessment of Earthquake Hazard in Sinai in Relation to the Seismicity in the Eastern Mediterranean Region. Bulletin of Engineering Geology and the Environment 2006; 65: 309-319.

[53] Abdel-Rahman K., Al-Amri AMS. and Abdel-Moneim E. Seismicity of Sinai Peninsula, Egypt. Arabian Journal of Geosciences 2009; 2 (2) 103-118.

[54] El-Hadidy M. Seismotectonics and Seismic Hazard Studies for Sinai Peninsula, Egypt. MSc. thesis. Ain Shams University, Egypt; 2008.

[55] El-Hadidy M. Seismotectonics and Seismic Hazard Studies in and around Egypt. PhD. thesis. Ain Shams University, Egypt; 2012.

[56] Deif A., Abou Elenean K., El-Hadidy M., Tealeb A. and Mohamed A. Probabilistic Seismic Hazard Maps for Sinai Peninsula, Egypt. Journal of Geophysics and Engineering 2009; 6: 288-297.

[57] Deif A., Hamed H., Igrahim HA., Abou Elenean K., and El-Amin EM. Seismic Hazard Assessment in Aswan, Egypt. Journal of Geophysics and Engineering 2011; 8: 531-548.

[58] Fat-Helbary RE., El Khashab HM., Dojcinovski D., El Faragawy KO., and Abdel-Motaal AM. Seismicity and Seismic Hazard Analysis in and around the Proposed Tushka New City Site, South Egypt. Acta Geodynamica et Geomaterialia 2008; 5 (4) 389-398.

[59] Mohamed AA., El-Hadidy M., Deif A., and Abou Elenean K. Seismic Hazard Studies in Egypt. NRIAG Journal of Astronomy and Geophysics 2012; 1: 119-140.

[60] Sawires R., Ibrahim HA., Fat-Helbary RE., and Peláez JA. A Seismological Database for Egypt Including Updated Seismic and Focal Mechanism Catalogues. 8th Spanish-

Portuguese Assembly of Geodesy and Geophysics. 29-31 January 2014, Évora, Portugal.

[61] Papazachos BC. Seismicity of the Aegean and Surrounding Area. Tectonophysics 1990; 178: 287-308.

[62] Shapira A., and Shamir G. Seismicity Parameters of Seismogenic Zones in and around Israel. The Institute of Petroleum Research and Geophysics 1994. Report No. Z1/567/79 (109).

[63] Papazachos BC. and Papaioannou ChA. Long–term Earthquake Prediction in the Aegean area Based on A Time and Magnitude Predicate Model. Pure Applied Geophysics 1993; 140: 595-612.

[64] Papioannou ChA. and Papazachos BC. Time–independent and Time–dependent Seismic Hazard in Greece Based on Seismogenic Sources. Bulletin of the Seismological Society of America 2000; 90: 22-33.

[65] CMT, Global Centroid Moment Tensor Catalogue: http://www.globalcmt.org/

[66] International Seismological Centre, On-line Bulletin, http://www.isc.ac.uk, International Seismological Centre, Thatcham, United Kingdom, 2011.

[67] PDE, Preliminary Determination of Epicentre: USGS National Earthquake Information Center (NEIC), http://earthquake.usgs.gov/earthquakes/.

[68] RCMT, European-Mediterranean RCMT Catalogue: http://www.bo.ingv.it/RCMT/.

[69] Said R. The Geology of Egypt. Elservier, Amsterdam; 1962.

[70] Said R. The Geological Evolution of the River Nile. Springer-Verlag, New York Inc., USA; 1981.

[71] Said R. The Geology of Egypt. A.A. Balkema, Ralkema, Rptterdam, Brookfield; 1990.

[72] Shata A. Structural Development of the Sinai Peninsula (Egypt). Conference Proceedings. 20th International Geological Congress, 1956, Mexico, 1959. p225-249.

[73] Neev D. Tectonic Evolution of the Middle East and the Levantine Basin (Easternmost Mediterranean). Geology 1975; 3: 683-686.

[74] Neev D., Almagor G., Arad A., Ginzburg A., and Hall J. The Geology of the Southern Mediterranean Sea. GSI, 68. 1976

[75] Neev D., and Hall JK. A Global System of Spiraling Geosutures. Journal of Geophysical Research 1982; 87:589-708.

[76] El-Shazly EM. The Geology of the Egyptian Region. In: The Ocean Basin and Margins. Volume 4A: The Eastern Mediterranean. Plenum Press. New York-London 1977.

[77] Maamoun M. Macroseismic Observation of Principal Earthquakes in Egypt. Bulletin of Helwan Institute of Astronomy and Geophysics. No. 183, 1979.

[78] Issawi B. Geology of the South Western Desert of Egypt. Annals of the Geological Survey of Egypt 1981; 2: 57-66.

[79] Riad S., EL-Etr HA., and Mokhles A. Basement Tectonics of Northern Egypt as Interpreted from Gravity Data. International Basement Tectonics Association Publication 1983; 4, 209-231.

[80] Sestini G. Tectonic and Sedimentary History of NE African Margin (Egypt/Libya). In: JE. Dixon and AHF. Robertson (eds.). The Geological Evolution of the Eastern Mediterranean. Blackwell Scientific Publications, Oxford, 1984. p161-175.

[81] Schlumberger. Geology of Egypt. Paper Presented at the Well Evaluation Conference, Schlumberger, Cairo; 1984.

[82] Woodward-Clyde Consultants. Earthquake Activity and Stability Evaluation for the Aswan High Dam. Unpublished report. High and Aswan Dam Authority, Ministry of Irrigation, Egypt; 1985.

[83] Barazangi M., Seber D., Chaimov T., Best J., Litak R., Al-Saad A and Sawaf T. Tectonic Evolution of the Northern Arabian Plate in Western Syria. In: Boschi E, Mantovani E and Morelli A. (ed.) Recent Evolution and Seismicity of the Mediterranean Region. Kluwer Academic Publishers, the Netherlands; 1993. p117–140.

[84] Guiraud RA. and Bosworth W. Phanerozoic Geodynamic Evolution of Northeastern Africa and the Northwestern Arabian Platform. Tectonophysics1999; 315: 73-108.

[85] Abdel Aal A., El Barkooky A., Gerrites M., Meyer H., Schwander M., and Zaki H. Tectonic Evolution of the Eastern Mediterranean Basin and its Significance for Hydrocarbon Prospectivity in the Ultradeep Water of the Nile Delta. The Mediterranean Offshore Conference. Alexandria, Egypt; 2000.

[86] Philobbos ER., Riad S., Omran AA. and Othman AB. Stages of Fracture Development Controlling the Evolution of the Nile Valley in Egypt. Egyptian Journal of Geology 2000; 44 (2) 503-532.

[87] Hussein IM. and Abd-Allah AM. Tectonic Evolution of the Northeastern Part of the African Continental Margin, Egypt. Journal of African Earth Sciences 2001; 33: 49-68.

[88] Drake CL., Girdler RW. A Geophysical Study of the Red Sea. Geophysical Journal of the Royal Astronomical Society 1964; 8: 473-495.

[89] Tramontini C. and Davies D. A Seismic Refraction Survey in the Red Sea. Geophysical Journal of the Royal Astronomical Society 1969; 17: 2225–2241.

[90] Tealeb A. Depth Determination of Density Contrasts in the Earth's Crust using Autocorrelation Analysis. Bulletin of Academy of Scientific Research, Helwan Institute of Astronomy and Geophysics, Cairo, No. 2008; 1979.

[91] Makris J., Stofen B., Maamoun M., Shehata W. Deep Seismic Sounding in Egypt, Part I: The Mediterranean Sea between Crete-Sidi Barani and the Coastal Area of Egypt. Unpublished Report, University of Hamburg ERG, 1979.

[92] Makris J., Allam A., Mokhtar T., Basahel A., Dehghani GA. and Bazari M. Crustal Structure in the Northwestern Region of the Arabian Shield and its Transition to the Red Sea. Bulletin of Faculty of Earth Sciences, King Abdulaziz University 1983; 6: 435-447.

[93] Makris J., Rihm R. and Allam A. Some Geophysical Aspects of the Evolution and Structure of the Crust in Egypt. In: Greiling SE.-G.a. R.O. (ed.) The Pan-African Belt of Northeast Africa and Adjacent Areas,Tectonic Evolution and Economic Aspects of a Late Proterozoic Orogen: Braunschweig, Friedr. Vieweg and Sohn; 1988. p345–369.

[94] Makris J., Henke CH., Egloff F. and Akamaluk T. The Gravity Field of the Red Sea and East Africa. Tectonophysics 1991; 198: 369–381.

[95] Rihm R. Seismische Messungen in Roten Meer und ihre interpretation. Diplom Arbeit, Institute fur Geophysik der Univeristat Hamburg, 1984.

[96] Gaulier JM., Le Pichon X., Lyberis N., Avedik F., Geli L., Moretti I., Deschamps A. and Hafez S. Seismic Study of the Crust of the Northern Red Sea and Gulf of Suez. Tectonophysics 1988; 153: 55–88.

[97] Rihm R., Makris J., Moller L. Seismic Surveys in the Northern Red Sea: Asymmetric Crustal Structure. Tectonophysics 1991; 198: 279–295.

[98] Shaaban MA., El Eraqi MA., Mamdouh EM. Deep Tectonics of Northern Eastern Desert of Egypt as Integrated from Gravity and Seismic Data. Journal of King Abdulaziz University: Earth Sciences 1994; 7: 75–88.

[99] Abd El-Hafiez H. The Role of Earthquake Analysis for Modeling the Dahshour Area Egypt. MSc. thesis. Ain Shams University, Egypt; 1996.

[100] Dorre AS., Carrara E., Cella F., Grimaldi M., Hady YA., Hassan H., Rapolla A. and Roberti N. Journal of African Earth Sciences 1997; 25: 425-434.

[101] Seber D., Steer D., Sandvol E., Sandvol C., Brindisi C., and Barazangi M. Design and Development of Information Systems for the Geosciences: An Application to the Middle East. GeoArabia 2000; 5 (2) 269-296.

[102] Mohamed H. and Miyashita K. One-dimensional Velocity Structure in the Northern Red Sea Area, Deduced from Travel Time Data. Earth's Planet Spaces 2001; 53: 695–702.

[103] El-Khrepy S. Tomographic Modeling of Dahshour Area Local Earthquakes, Northern Egypt. MSc. thesis. Ain Shams University, Egypt, 2001.

[104] El-Khrepy S. Detailed Study of the Seismic Waves Velocity and Attenuation Models Using Local Earthquakes in the Northeastern Part of Egypt. PhD. thesis. Mansoura University, Egypt, 2008.

[105] Koulakov I. and Sobolev SV. Moho Depth and Three-dimensional P and S Structure of the Crust and Uppermost Mantle in the Eastern Mediterranean and Middle East Derived from Tomographic Inversion of Local ISC Data. Geophysical Journal International 2006; 164: 218-235.

[106] Gharib A. Crustal Structure of Tushka Region, Abu-Simbel, Egypt, Inferred from Spectral Ratios of P-waves of Local Earthquakes. Acta Geophysica 2006; 54: 361- 377.

[107] Salah MK. Crustal Structure Beneath Kottamiya Broadband Station, Northern Egypt from Analysis of Teleseismic Receiver Functions. Journal of African Earth Sciences 2011; 60: 353–362.

[108] Abdelwahed MF., El-Khrepy S., and Qaddah A. Three-dimensional Structure of Conrad and Moho Discontinuities in Egypt. Journal of African Earth Sciences 2013; 85: 87-102.

[109] Araya R., and Der Kiureghian A. Seismic Hazard Analysis: Improved Models, Uncertanities and Sensitivities. Report to the National Science Foundation, Earthquake Enginnering Research Center 1988. Report No. UCB/EERC-90/11.

[110] Gutenberg B. and Richter CF. Frequency of Earthquakes in California. Bulletin of the Seismological Society of America1944; 34: 185-188.

[111] Salamon A., Hofstetter A., Garfunkel Z. and Ron H. Seismicity of the Eastern Mediterranean Region: Perspective from the Sinai Subplate. Tectonophysics 1996; 263: 293-305.

[112] Salamon A., Hofstetter A., Garfunkel Z. and Ron H. Seismotectonics of the Sinai Subplate-the Eastern Mediterranean Region. Geophysical Journal International 2003; 155: 149-173.

[113] Garfunkel Z., Zak I. and Freund R. Active Faulting in the Dead Sea Rift. Tectonophysics 1981; 80: 1–26.

[114] Reches Z. and Hoexter DF. Holocene Seismic Activity in the Dead Sea Area. Tectonophysics 1981; 80: 235-254.

[115] Marco S., Agnon A., Ellenblum R., Eidekman A., Basson U. and Boas A. 817-year-old Walls Offset Sinistrally 2.1 m by the Dead Sea Transform, Israel. Journal of Geodynamics 1997; 24:11–20.

[116] Marco S., Heimann A., Rockwell KT. and Agnon A. Late Holocene Earthquake Deformations in the Jordan Gorge Fault, Dead Sea Transform. In Abstracts of Israel Geological Society Annual Meeting. Ma'alot, 2000, p85.

[117] Zilberman E., Amit R., Heimann A. and Porat N. Changes in Holocene Paleoseismic Activity in the Hula Pull-apart Basin, Dead Sea Rift, Northern Israel. Tectonophysics 2000; 321: 237–252.

[118] Amit R., Zilberman E., Porat N., Enzel Y. Relief Inversion in the Avrona Playa as Evidence of Large-magnitude Historical Earthquakes, Southern Arava Valley, Dead Sea Rift. Quaternary Research 1999; 52 (1) 76 – 91.

[119] Zhang H. and Niemi TM. Slip Rate of the Northern Wadi Araba Fault, Dead Sea Transform, Jordan. GSA Abstracts with Programs 1999; 31, A-114.

[120] Klinger Y., Avouac J., Abou Karaki N., Dorbath L., Bourles and Reyss J. Slip Rate on the Dead Sea Transform Fault in Northern Araba Valley, Jordan. Geophysical Journal International 2000; 142: 755-768.

[121] Gomez F., Meghraoui M., Darkal AN., Sbeinati R., Darawcheh R., Tabet C., Knawlie M., Charabe M., Khair K. and Barazangi M. Coseismic Displacements along the Serghaya Fault: An Active Branch of the Dead Sea Fault System in Syria and Lebanon. Journal of the Geological Society of London 2001; 158: 405-408.

[122] Garfunkel Z. Internal Structure of the Dead Sea Leaky Transform (Rift) in Relation to Plate Kinematics. Tectonophysics 1981; 80, 81-108.

[123] Walley CD. A Braided Strike-slip Model for the Northern Continuation of the Dead Sea Fault, and its Implications to Levantine Tectonics. Tectonophysics 1988; 145: 63–72.

[124] Girdler RW. The Dead Sea Transform Fault System. Tectonophysics 1990; 180: 1–13.

[125] Gomez F., Meghraoui M., Darkal AN., Hijazi F., Mouty M., Suleiman Y., Sbeinati R., Darawcheh R., Al-Ghazzi R., Barazangi M. Holocene Faulting and Earthquake Recurrence along the Serghaya Branch of the Dead Sea Fault System in Syria and Lebanon. Geophysical Journal International 2003; 153: 1–17.

[126] Meghraoui M., Cakir Z., Masson F., Mahmoud Y., Ergintav S., Alchalbi A., Inan S., Daoud M., Yonlu O., Altunel E. Kinematic Modelling at the Triple Junction between the Anatolian, Arabian, African Plates (NW Syria and in SE Turkey). Geophysical Research Abstracts 2011; 13, EGU2011-12599, EGU General Assembly, Vienna.

[127] Karabacak V., and Altunel E. Evolution of the Northern Dead Sea Fault Zone in Southern Turkey. Journal of Geodynamics 2013; 65: 282–291.

[128] Mahmoud Y., Masson F., Meghraoui M., Cakir Z., Alchalbi A., Yavasoglu H., Yönlü O., Daoud M., Ergintav S., Inan S. Kinematic Study at the Junction of the East Anatolian Fault and the Dead Sea Fault from GPS Measurements. Journal of Geodynamics 2013; 67: 30–39.

[129] Dziewonski AM., Ekstrom G., Salganik MP. Centroid Moment Tensor Solutions for October – December 1995. Physics of the Earth and Planetary Interiors 1997; 101: 1 – 12.

[130] Garfunkel Z. The Tectonics of the Western Margins of the South Arava. PhD. thesis. The Hebrew University of Jerusalem, Israel, 1974.

[131] Ben-Avraham Z., Almagor G. and Garfunkel Z. Sediments and Structure of the Gulf of Elat (Aqaba) Northern Red Sea. Sedimentary Geology 1979; 23: 239-267.

[132] Eyal M, Eyal Y., Bartov Y. and Steinitz G. The Tectonic Development of the Western Margin of the Gulf of Eilat (Aqaba) Rift. Tectonophysics 1981, 80: 39-66.

[133] Shapira A., Jarradat M. Earthquake Risk and Loss Assessment in Aqaba and Eilat Regions. Submitted to the US Aid-Merc Program 1995.

[134] Heidbach O., and Ben-Avraham Z. Stress Evolution and Seismic Hazard of the Dead Sea Fault System. Earth and Planetary Science Letters 2007; 257: (1-2) 299-312.

[135] Pinar A. and Türkelli N. Source Inversion of the 1993 and 1995 Gulf of Aqaba Earthquakes. Tectonophysics 1997; 283: 279–288.

[136] Klinger Y., Rivera L., Haessler H. and Maurin JC. Active Faulting in the Gulf of Aqaba: New Knowledge from the M_W 7.3 Earthquake of 22 November 1995. Bulletin of Seismological Society of America 1999; 89: 1025–1036.

[137] Hofstetter A., Thio HK. and Shamir G. Source Mechanism of the 22/11/1995 Gulf of Aqaba Earthquake and its Aftershock Sequence. Journal of Seismology 2003; 7: 99-114.

[138] Abdel Fattah AK., Hussein HM. and El Hady S. Another Look at the 1993 and 1995 Gulf of Aqaba Earthquake from the Analysis of Teleseismic Waveforms. Acta Geophysica 2006; 54 (3) 260-279.

[139] Lyberis N. Tectonic Evolution of the Gulf of Suez and the Gulf of Aqaba. Tectonophyics 1988; 153: 209-220.

[140] Bayer H., Hötzl H., Jado A., Bocher B. and Voggenreiter W. Sedimentary and Structural Evolution of the Northwest Arabian Red Sea Margin. Tectonophysics 1988; 153: 137-151.

[141] Gerson R., Grossman S., Amit R. and Greenbaum N. Indicators of Faulting Events and Periods of Quiescence in Desert Alluvial Fans. Earth Surface Processes and Landforms 1993; 18: 181-202.

[142] Steinitz G., Bartov Y., and Hunziker JC. K-Ar Age Determinations of Some Miocene-Pliocene Basalts in Israel: Their Significance to the Tectonics of the Rift Valley. Geological Magazine 1978; 115: 329–340.

[143] Moustafa AR. and Khalil MH. Superposed Deformation in the Northern Suez Rift, Egypt: Relevance to Hydrocarbon Exploration. Journal of Petroleum Geology 1995; 18: 245–266.

[144] Bartov Y. A Structural and Paleographic Study of the Central Sinai Faults and Domes. PhD. thesis (in Hebrew with an English Abstract). The Hebrew University of Jerusalem, Israel; 1974.

[145] Moustafa AR. and Khalil MH. Rejuvenation of the Eastern Mediterranean Passive Continental Margin in Northern and Central Sinai: New Data from the Themed Fault. Geological Magazine 1994; 131: 435-448.

[146] El-Isa Z., Merghelani H. and Bazari M. The Gulf of Aqaba Earthquake Swarm of 1983. Geophysical Journal of the Royal Astronomical Society 1984; 76: 711-722.

[147] Ben-Menahem A., Nur A., and Vered M. Tectonics, Seismicity and Structure of the Afro-Eurasian Junction--The Breaking of an Incoherent Plate. Physics of the Earth and Planetary Interiors 1976; 12: 1-50.

[148] Gardosh M., Reches Z. and Garfunkel Z. Holocene Tectonic Deformation along the Western Margins of the Dead Sea. Tectonophysics 1990; 180: 132-137.

[149] Heimann A. The Development of the Dead Sea Rift and its Margins in Northern Israel in the Pliocene and Pleistocene. PhD. thesis. The Hebrew University of Jerusalem, Israel; 1990.

[150] Shamir G., Bartov A., Fleischer L., Arad V., Rosensaft M. Preliminary Seismic Zonation. Geological Survey of Israel. Report No. GSI/12/2001, Geophysical Institute of Israel Report No. GII 550/ 95/01/ (1); 2001.

[151] Cochran JR. A Model for the Development of the Red Sea. American Association of Petroleum Geologists Bulletin 1983; 67: 41-69.

[152] Cochran JR., Martinez F., Steckler and Hobart MA. Conrad Deep, a New Northern Red Sea Deep, Origin and Implications for Continental Rifting. Earth and Planetary Science Letters 1986; 78: 18-32.

[153] Girdler RW and Styles P. Two Stage Sea-floor Spreading. Nature1974; 247: 7-11.

[154] Roeser HA. A Detailed Magnetic Survey of the Southern Red Sea. Geologie Jahrbuch 1975; 13: 131–153.

[155] LaBrecque JL., and Zitellini N. Continuous Sea Floor Spreading in the Red Sea: An Alternative Interpretation of Magnetic Anomaly Pattern. The American Association of Petroleum Geologists Bulletin 1985; 4: 513-524.

[156] Makris J. and Rihm R. Shear-controlled Evolution of the Red Sea: Pull-Apart Model. Tectonophysics 1991; 198: 441–466.

[157] Searle RC. and Ross DA. A Geophysical Study of the Red Sea Axial Trough Between 20.58 and 22.8N. Geophysical Journal of the Royal Astronomical Society 1975; 43: 555–572.

[158] Ghebreab W. Tectonics of the Red Sea Region: Reassessed. Earth-Science Reviews 1998; 45: 1-44.

[159] Cochran JR. and Martinez F. Evidence from the Northern Red Sea on the Transition from Continental to Oceanic Rifting. Tectonophysics 1988; 153: 25-53.

[160] Bonati E. Punctiform Initiation of Seafloor Spreading in the Red Sea during Transition from a Continental to an Oceanic Rift. Nature 1985; 316: 7-33.

[161] Daggett P., Morgan P., Boulous F., Hennin S., El-Sherif A., El-Sayed A., Basta N. and Melek Y. Seismicity and Active Tectonics of the Egyptian Red Sea Margin and the Northern Red Sea. Tectonophysics 1986; 125: 313-324.

[162] McKenzie DP., Davies D. and Molnar P. Plate Tectonics of the Red Sea and East Africa. Nature 1970; 226: 243–248.

[163] Robson D. The Structure of the Gulf of Suez (Clysmic) Rift with Special Reference to the Eastern Side. Geological Society of London 1971; 115: 247-276.

[164] Garfunkel Z. and Bartov Y. The Tectonics of the Suez Rift. Geological Survey of Israel Bulletin 1977.

[165] Jackson J., White N., Garfunkel Z. and Anderson H. Relation between Normal-fault Geometry, Tilting and Vertical Motions in Extensional Terrains: An Example from the Southern Gulf of Suez. Journal of Structural Geology 1988; 10 (2) 155-170.

[166] Bosworth W. and Taviani M. Late Quaternary Reorientation of Stress Field and the Extension Direction in the Southern Gulf of Suez, Egypt: Evidence from Uplifted Coral Terraces, Mesoscopic Fault Arrays and Borehole Breakouts. Tectonics 1996; 15: 791-802.

[167] Moustafa AR. Internal Structure and Deformation of an Accommodation Zone in the Northern Part of the Suez Rift. Journal of Structural Geology 1996; 18: 93-107.

[168] Moustafa AM. Block Faulting in the Gulf of Suez. Conference Proceedings, 1976, Cairo, Egypt. 5th Egyptian General Petroleum Corporation Exploration Seminar, 35 p.

[169] Bosworth W. A High-strain Rift Model for the Southern Gulf of Suez, Egypt. In: Lambiase L L (ed.) Hydrocarbon Habitat in Rift Basins. Geological Society of London 1985. Special Publication 80, p75–102.

[170] Moustafa AR. and Fouda HG. Gebel Sufr El Dara Accommodation Zone, Southwestern Part of the Suez rift. Middle East Research Center, Ain Shams University, Earth Science Series 1988; 2: 227–239.

[171] Chorowicz J. and Sorlien C. Oblique Extensional Tectonics in the Malawi Rift, Africa. Geological Society of America Bulletin 1992; 104: 1015–1023.

[172] Maler MO. Dead Horse Graben: A West Texas Accommodation Zone. Tectonics 1990; 9: 1257–1268.

[173] Boccaletti M., Getaneh A., and Tortoorici L. The Main Ethiopian Rift: An Example of Oblique Rifting. Annales Tectonicae 1992; 6: 20–25.

[174] Lacombe O., Angelier J., Byrne D., and Dupin JM. Eocene-Oligocene Tectonics and Kinematics of the Rhine-Saone Continental Transform Zone (Eastern France). Tectonics 1993; 12: 874-888.

[175] Colletta B., Le Quellec P., Letouzey J., and Moretti I. Longitudinal Evolution of the Suez Rift Structure, Egypt. Tectonophysics 1988; 153: 221–233.

[176] Le Pichon X. and Gaulier J. The Rotation of the Arabia and Levant Fault System. Tectonophysics 1988; 153: 271-294.

[177] Mart Y. The Dead Sea Rift: From Continental Rift to Incipient Ocean. Tectonophysics 1991; 197:155–179.

[178] Ben-Menahem A. Earthquake Catalogue for the Middle East (92 BC-1980 AD). Bollettino di Geofisica Teorica e Applicata 1979; 21: 245-310.

[179] Ben-Menahem A. and Aboodi E. Tectonic Pattern in the Northern Red Sea Region. Journal of Geophysical Research 1971; 76: 2674-2689.

[180] Younes AI., and McClay K. Development of Accommodation Zones in the Gulf of Suez-Red Sea Rift, Egypt. The American Association of Petroleum Geologists Bulletin 2002; 86: 1003-1026.

[181] Fairhead JD. and Girdler RW. The Seismicity of the Red Sea, Gulf of Aden and Afar Triangle. Philosophical Transactions of the Royal Society of London 1970; 267: 49-74.

[182] Maamoun M. and El Khashab HM. Seismic Studies of the Shedwan (Red Sea) Earthquake. Helwan Institute of Astronomy and Geophysics. No. 171, 1978.

[183] Jackson JA. and McKenzei DP. Active Tectonics of the Alpine Himalayan Belt between Western Turkey and Pakistan. Geophysical Journal of the Royal Astronomical Society 1984; 77: 185-264.

[184] Huang P. and Soloman S. Centroid Depth and Mechanisms of Mid-ocean Ridge. Journal of Geological Research 1987; 92: 1361-1383.

[185] Badawy A., and Horváth F. Sinai Subplate and Kinematic Evolution of the Northern Red Sea. Journal of Geodynamics 1999a; 27: 433-450.

[186] Badawy A., and Horváth F. Seismicity of the Sinai Subplate Region: Kinematic Implications. Journal of Geodynamics 1999b; 27: 451-468.

[187] Badawy A., and Horváth F. Recent Stress Field of the Sinai Subplate Region. Tectonophysics 1999c; 304: 385-403.

[188] Badawy A. Status of the Crustal Stress as Inferred from Earthquake Focal Mechanisms and Borehole Breakouts in Egypt. Tectonophysics 2001; 343 (1-2) 49-61.

[189] El-Gaby S. Architecture of the Egyptian Basement Complex. 5th International Conference on Basement Tectonics, Egypt; 1983.

[190] El-Gaby S., List FK., and Tehrani R. Geology, Evolution and Metallogenesis of the Pan-African Belt in Egypt. In: S.El-Gaby and R. O. Greiling (eds), The Pan-African Belt of Northeast African and Adjacent Areas, Fried. Vieweg and Shon, Braun Schweig, Wiesbaden 1988, p17-68.

[191] Stern RJ., and Hedge CE. Geochronologic and Isotopic Constraints on Late Precambrian Crustal Evolution in the Eastern Desert of Egypt. American Journal of Science 1985; 285: 97-127.

[192] Issawi B. The Geology of Kurkur-Dungul Area. General Egyptian Organization for Geological Research and Mining; Cairo, Egypt. Geological Survey. No. 46, 101 pp; 1969.

[193] Issawi B. Geology of the Southwestern Desert of Egypt. Annals of the Geological Survey of Egypt 1982, 215 p.

[194] Issawi B. Geology of the Aswan Desert. Annals of the Geological Survey of Egypt; 1987.

[195] Fat-Helbary RE. A Study of the Local Earthquake Magnitude Determination Recorded by Aswan Seismic Network. MSc. thesis. Assiut University, Sohag Branch, Egypt; 1989.

[196] El-Younsy ARM., Ibrahim HA., Senosy MM. and Galal WF. Structural Characteristics and Tectonic Evolution of the Area around the Qena Bend, Middle Egypt. 6th International Conference on the Geology of Africa, 2010, Assiut, Egypt.

[197] Beadnell HJL. Dakhla Oasis, Its Topography and Geology: Egypt. Survey Department 1901; 104 Pages, 9 Maps, 7 Figures.

[198] Sandford KS. Paleolithic Man and the Nile Valley in Upper and Middle Egypt. Chicago University Oriental Institute Publications 1934; 18: 1-131.

[199] El-Gamili M. A Geophysical Interpretation of A part of the Nile Valley, Egypt Based on Gravity Data. Journal of Geology 1982. Special Volume, Part 2: 101-120.

[200] Abdel-Rahman MA., and El-Etr HA. The Orientational Characteristics of the Structure Grain of the Eastern Desert of Egypt. In: Symposium of the Evolution and Mineralization of the Arabian-Nubian Shield. Institute of Applied Geology, Jeddah, Saudi Arabia; 1978.

[201] Abdel Tawab S., Helal A., Deweidar H. and El-Sayed A. Surface Tectonic Features of 12 Oct., 1992 Earthquake, Egypt, at the Epicentral area. Ain Shams Scientific Bulletin 1993; Special Issue, p124-136.

[202] Maamoun M., Megahed A., Hussein A. and Marzouk I. Preliminary Studies on Dahashour Earthquake. National Research Institute of Astronomy and Geophysics, Cairo, Egypt. (Abstract), 1993.

[203] Mousa HH. Earthquake Activity in Egypt and Adjacent Regions and its Relation to the Geotectonic Features. MSc. thesis. Mansoura University, Egypt, 1989.

[204] Hassib GH. A Study of Focal Mechanism for Recent Earthquakes in Egypt and their Tectonic Implication. MSc thesis. Assiut University, Egypt, 1990.

[205] Megahed A. and Dessoky MM. The Ismailia (Egypt) Earthquake of January 2nd, 1987 (Location, Macroseismic Survey, Radiation Pattern of First Motion and its Tectonic Implications), 1988.

[206] Hassoup A. and Tealab A. Attenuation of Intensity in the Northern Part of Egypt Associated with the May 28, 1998 Mediterranean Earthquake. Acta Geophysica Polonica 2000; 48: 79-92.

[207] Riad S. and Hosney H. Fault Plane Solution for the Gilf Kebir Earthquake and the Tectonics of the Southern Part of the Western Desert of Egypt. Annals of the Geological Survey of Egypt 1992; 18: 239-248.

[208] Sykes LR. Intraplate Seismicity, Reactivation of Preexisting Zones of Weakness, Alkaine Magmatism and Other Tectonic Postdating Continental Fragmentation. Reviews of Geophysics and Space Physics 1987; 16: 621-688.

[209] Talwani P. and Rajendrank. Some Seismological and Geometric Features of Intraplate Earthquakes. Seismological Research Letters 1991; 59, 305-310.

[210] McKenzie D. Plate Tectonics of the Mediterranean Region. Nature1970; 326: 239-243.

[211] McKenzie D. Active Tectonics in the Mediterranean Region. Geophysical Journal of the Royal Astronomical Society 1972; 30: 109-185.

[212] Dewey JF., Pitman WC., Ryan WBF. and Bonnin J. Plate Tectonics and the Evolution of the Alpine System. Bulletin of the Geological Society of America 1973; 84: 3137-3180.

[213] Westaway R. Present-day Kinematics of the Middle East and Eastern Mediterranean. Journal of Geophysical Research 1994; 99 (6) 12071-12090.

[214] Kiratzi A. and Papazachos C. Active Crustal Deformation from the Azores Triple Junction to the Middle East. Tectonophysics 1995; 243: 1-24.

[215] Ambraseys N. and Jackson J. Faulting Associated with Historical and Recent Earthquakes in the Eastern Mediterranean Region. Geophysical Journal International 1998; 133: 390-406.

[216] McClusky S., Balassanian S., Barka A., Demir C., Ergintav S., Georgiev I., Gurkan O., Hamburger M., Hurst K., Kahle H., Kastens K., Kekelidze G., King R., Kotzev V., Lenk O., Mahmoud S., Mishin A., Nadariya M., Ouzounis A., Paradissis D., Peter Y., Prilepin M., Reilinger R., Sanli I., Seeger H., Tealeb A., Toksoz MN. and Veis G. Global Positioning System Constraints on Plate Kinematics and Dynamics in the Eastern Mediterranean and Caucasus. Journal of Geophysical Research 2000; 105 (3) 5695–5719.

[217] Vidal N., Alvarez-Marron J. and Klaeschen D. Internal Configuration of the Levantine from Seismic Reflection Data (Eastern Mediterranean). Earth and Planetary Science Letters 2000a; 180: 77–89.

[218] Vidal N., Alvarez-Marron J. and Klaeschen D. The Structure of the Africa–Anatolia Plate Boundary in the Eastern Mediterranean. Tectonics 2000b, 19: 723–739.

[219] Vidal N., Klaeschen D., Kopf A., Docherty C., Von-Huene R. and Krasheninnikov VA. Seismic Images at the Convergence Zone from South of Cyprus to the Syrian Coast, Eastern Mediterranean. Tectonophysics 2000c; 329: 157–170.

[220] Papazachos BC. and Papaioannou ChA. Lithospheric Boundaries and Plate Motions in the Cyprus Area. Tectonophysics 1999; 308: 193–204.

[221] Ziegler PA. Evolution of the Arctic-North Atlantic and Western Tethys. The American Association of Petroleum Geologists Memoir 1988; 43, 198p.

[222] Meulenkamp JE., Wortel MJR., van Wamel WA., Spakman W. and Hoogerduyn Strating E. On the Hellenic Subduction Zone and the Geodynamic Evolution of Crete since the Late Middle Miocene. Tectonophysics 1988; 146: 203–215.

[223] Dewey JF., Helman ML., Turco E., Hutton DHW., and Knott SD. Kinematics of the Western Mediterranean. In: Alpine Tectonics, edited by M.P. Coward, D. Detrich and R.G. Park, Geological Society of London, Special Publication 1989, 45: 265-283.

[224] SHARE "Seismic Hazard Harmonization in Europe", 2013. http://www.share-eu.org/.

[225] Woessner J., Giardini D., and the SHARE consortium. Seismic Hazard Estimates for the Euro-Mediterranean Region: A Community-based Probabilistic Seismic Hazard Assessment. Proceedings of the 15th World Conference of Earthquake Engineering, Lisbon, Portugal, 2012.

[226] Papioannou ChA. A Model for the Shallow and Intermediate-depth Seismic Sources in the Eastern Mediterranean Region. Bollettino di Geofisica 2001; 42: 57-73.

[227] Demircioglu MB., Sesetyan K., Durukal E. and Erdik M. Assesement of Earthquake Hazard in Turkey. Conference Proceedings, 4th International Conference on Earthquake Geotechnical Engineering, 25-28 June 2007, Thessaloniki, Greece. Springer, New York.

[228] Demircioglu MB. The Earthquake Hazard and Risk Assessment for Turkey. PhD. thesis. Bogazici University, Turkey, 2010.

[229] Papazachos BC., and Comninakis PE. Geophysical Features of the Greek Island Arc and Eastern Mediterranean Ridge. Final Proceedings. Seances de la Conference Reunie a Madrid, 1969/1970. Madrid, Spain. 16: 74-75.

[230] Papazachos BC., and Comninakis PE. Geophysical and Tectonic Features of the Aegean Arc. Journal of Geophysical Research 1971; 76: 8517-8533.

[231] Comninakis PE., and Papazachos BC. Space and Time Distribution of the Intermediate Depth Earthquakes in the Hellenic Arc. Tectonophysics 1980; 70: 35-47.

[232] Papazachos BC., Papadimitriou EE., Karakostas BG. and Karakaisis GF. Long-term Prediction of Great Intermediate-Depth Earthquakes in Greece. Proceedings of the 12th Regional Seminar on Earthquake Engineering, EAEE-EPPO, Halkidiki, 1985, 1-12.

9

Rotational Components of the Seismic Fields Caused by Local Events

Anna Kurzych, Krzysztof P. Teisseyre,
Zbigniew Krajewski and Leszek R. Jaroszewicz

1. Introduction

The existence of rotation effects at the Earth surface associated with earthquakes has been observed probably at least from the times when scientific approach to the ground motions during the quake had started. They are described in several classical monographs, such as Hobbs [1] and Davison [2], in which cited examples concern, among other things, twisting of some obelisks, tombs and segments of columns. However, early publications explain such phenomena as incidental effects of interference between linear vibrations [3, 4]. For instance, Imamura [5] explained the rotation effects of some objects at the ground surface by the impact of body/surface waves: due to such impact, an object can be inclined, partly losing contact with the ground surface, and when returning to the vertical some twist occurs with respect to its former position. Hence, from the beginning, the rotational effects have been treated as derivative effects, and it was stated that although such effects are observed, they cannot be explained as effect of any rotational waves - or rotational components of seismic waves - because existence of such waves or components would contradict the ideal elastic theory [6].

In the second half of last century, it was observed a spectacular development of continuum mechanics including defects, granular structure and other deviations from the ideal linear elasticity. Special interests were concentrated on the micropolar and micromorphic continua. In such elastic continua, the real rotation can be accompanied by other kind of axial motion – the twist-bend motion. On above base, it was theoretically proved that so-called the seismic rotation waves could propagate through grained rocks, initially by Teisseyre [7] who initially attributed the appearance of rotation components in seismic wave by coupling the seismic waves with the micromorphic response of the medium characterized by the an internal/ granular structure [7, 8]. From this time, this possibility was extended to rocks with microstructure or defects [9, 10] or even without any internal structure [11 -13], due to the asymmetric stresses in the medium. On this base, various types of rotational waves have been discussed theoretically [14, 15].

It should be stressed that seismologists share different opinions about the nature of rotation waves – see the preface of a monograph on rotational seismology [16]. Perhaps, as it is underlined in preface of a book [16], still the majority believes that such rotation motions are not related to inner rotations but are directly related to rotation in the displacement field which may reach much higher magnitudes in materials with an internal structure than in homogenous layers; considering damages in the high buildings, there are many examples indicating enormous increase of rotation effect caused by consecutive impacts of seismic body and surface waves.

Nevertheless, all above aspects can be treated as elements of rotational seismology. It is an emerging field for studying all aspects of rotational ground motions induced by earthquakes, explosions, an ambient vibrations. It should be noticed that nowadays there is observed rapid growth of the rotational seismology interest in many geophysical fields of knowledge [17] which includes wide seismology disciplines, seismological apparatus, seismic-origin phenomena, physical and engineering aspects of earthquakes as well as geodesy.

However practical aspect of rotational events and phenomena investigation is connected with method of their recording, and different rotational seismology branches need different devices. For example, earthquake physics need devices operating with sensitivity below 10^{-9} rad/s/$Hz^{1/2}$, whereas the engineering of a strong-motion seismology devices operating with a frequency range 0.05-100 Hz with sensitivity 10^{-6}-10^{-1} rad/s/$Hz^{1/2}$ [18]. In this subject, it should be noticed that the seismic rotation waves were for the first time effectively recorded in Poland in 1976 [19]. Even though, from this time, waves or phenomena of this type have been studied in a few centers over the world, a further experimental verification of the existing rotational phenomena needs a new approach to the construction of the measuring devices, because the conventional seismometers are inertial sensors detecting only linear velocities [20]. Thus, during measurement of the rotation present in the seismic field, with the use of a special array or set of conventional seismometers (for example based on a set of two classical mechanical seismometers [21]), data are disturbed by linear movements [22]. Therefore, an innovative device is necessary to detect the rotational seismic phenomena/events. According to our knowledge we can confirm that the technical implementation of the Saganc effect [23] is the most proper way to measure rotation directly. One can find instances of such solution: a ring laser [24] as well as a fibre optic seismometer [25-27]. It gives the opportunity to carry out the measurements without any reference system.

It should be underlined that all experimental data recorded during earthquakes shows that rotational components are small in comparison to linear motions - less than 10% [14, 19] or have half of above value and exist with some delay regarding last one [25].

Based on above review in this chapter we present an analysis of a few examples of the rotational seismograms. Authors have concentrated on the local seismic events obtained at the Książ Observatory in Poland. These signals were obtained from two kinds of sensors described in section 2: the micro-array of TAPSs – Twin Antiparallel Pendulum Seismometers (also named rotation seismometers or double pendulum horizontal electromagnetic seismometers) and with the Sagnac interferometer of AFORS - Autonomous Fibre-Optic Rotational Seismometer, constructed at the Institute of Geophysics and the Military University of Technology, respectively. It should be also underlined that signals derived from micro-array include two components which, according to Asymmetric Continuum Theory, have character of rotational wave: rotation ω and shear E (called also pure shear) – see [14]. On the other side, the Sagnac-type seismometers detect only rotation and are completely insensitive to translations [25, 28, 29] which may contaminate rotational measurements. Nevertheless, probably all the signals analyzed here suffer from some disturbances, this is referred to in the section 3.

2. Instrumentation for recording rotational components of the seismic events

Figure 1 presents the general view of the measurement devices installed, at the beginning of July 2010, in the Książ Observatory, Poland (located at 50.84380333N, 16.291755 E). There are AFORS-1, the micro-array of seismometers consisting of two TAPSs (TAPS-1 and TAPS-2) oriented perpendicularly in the N-S and E-W directions, and other instruments such as accelerometers (parallel positioned with TAPSs), etc.

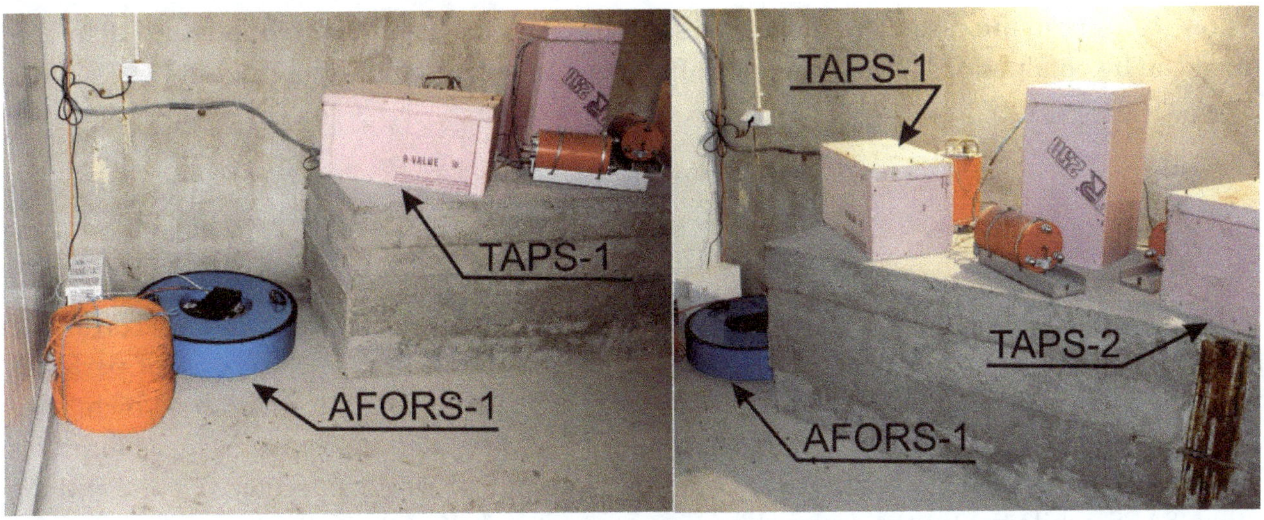

Figure 1. The general view of measurement devices installed in the Książ Observatory

The data detected by TAPSs (two channels for each of them) are stored by standard seismological system KST while data detected by AFORS are stored both by FORS-Telemetric Server and KST. The KST system uses sampling of the signals with frequency equals 12,8 kHz. The process of data storing by KST uses frequency of 100 Hz. Figure 2 presents an example of a diagram with data collected on March 11th, 2011 at 6 h 58 min. (after the Honshu earthquake

M=9.0 on March 11th, 2011 at 5 h 46 min. 23 s UT, recorded in Książ, Poland on March 11th, 2011 at 5 h 58 min. 35 s UT), used in previously presented analysis [30].

2.1. Design of the Twin Antiparallel Pendulum Seismometers

The micro-array of seismometers (system of two TAPSs perpendicularly oriented) is an experimental apparatus, devised in the Institute of Geophysics, and manufactured according to description presented below, on the base of short period SM-3 seismometers. This is one of the simplest micro-arrays for measuring the rotation and twist (shear) [14]. It was deployed at two Polish observatories, in Książ and Ojców (see [31]). The third identical set of sensors was used in Central Italy [32].

The idea of using the classical short period SM-3 seismometer as a new kind of mechanical rotational seismometer named TAPS is presented in Figure 3 [21]. It is a set of two SM-3 seismometers (named in Figure 3b as left – L and right – R) situated on a common axis and connected in parallel, but with opposite orientation. In the case of the ground motion containing displacements $w(t)$ and rotation $\alpha(t)$, the $u(t)$ - electromotive force recorded by each simple seismometer contains a component of displacement ($\pm w$) and rotation motion (α) multiplied by a proper length of pendulum (l) [33]:

$$u(t)_{L,\,R} = \pm w\left(t\right) + l \cdot \alpha(t), \tag{1}$$

where sign "+" and "-" are for R and L seismometer, respectively.

As one can see, in the case of identical two seismometers the rotational motions and displacement can be obtained from the sum and difference of the two recorded signals as:

$$\begin{aligned} \alpha(t) &= [u(t)_R + u(t)_L] / 2l, \quad \text{a} \\ w\left(t\right) &= [u(t)_R - u(t)_L] / 2l. \quad \text{b} \end{aligned} \tag{2}$$

If the ground could be treated as a perfect rigid body, then the rotational motion recorded by sole one TAPS was identical to rotation. But rocks and the ground surface are not perfectly rigid; they transfer the mechanical waves due to slight, transient deformations which, seen along different axes, may different. Consequently, rotation ω in the plane of measurements and the given moment are calculated as a mean of rotational motions received at one and the other TAPS, while the twist E (pure shear) is obtained as half of their difference:

$$\begin{aligned} \omega(t) &= [\alpha_1(t) + \alpha_2(t)] / 2, \quad \text{a} \\ E(t) &= [\alpha_1(t) - \alpha_2(t)] / 2. \quad \text{b} \end{aligned} \tag{3}$$

Relations (2a) and (2b) remain valid, however, only when both seismometers forming the system have exactly the same response characteristics. Because, as a matter of fact, the

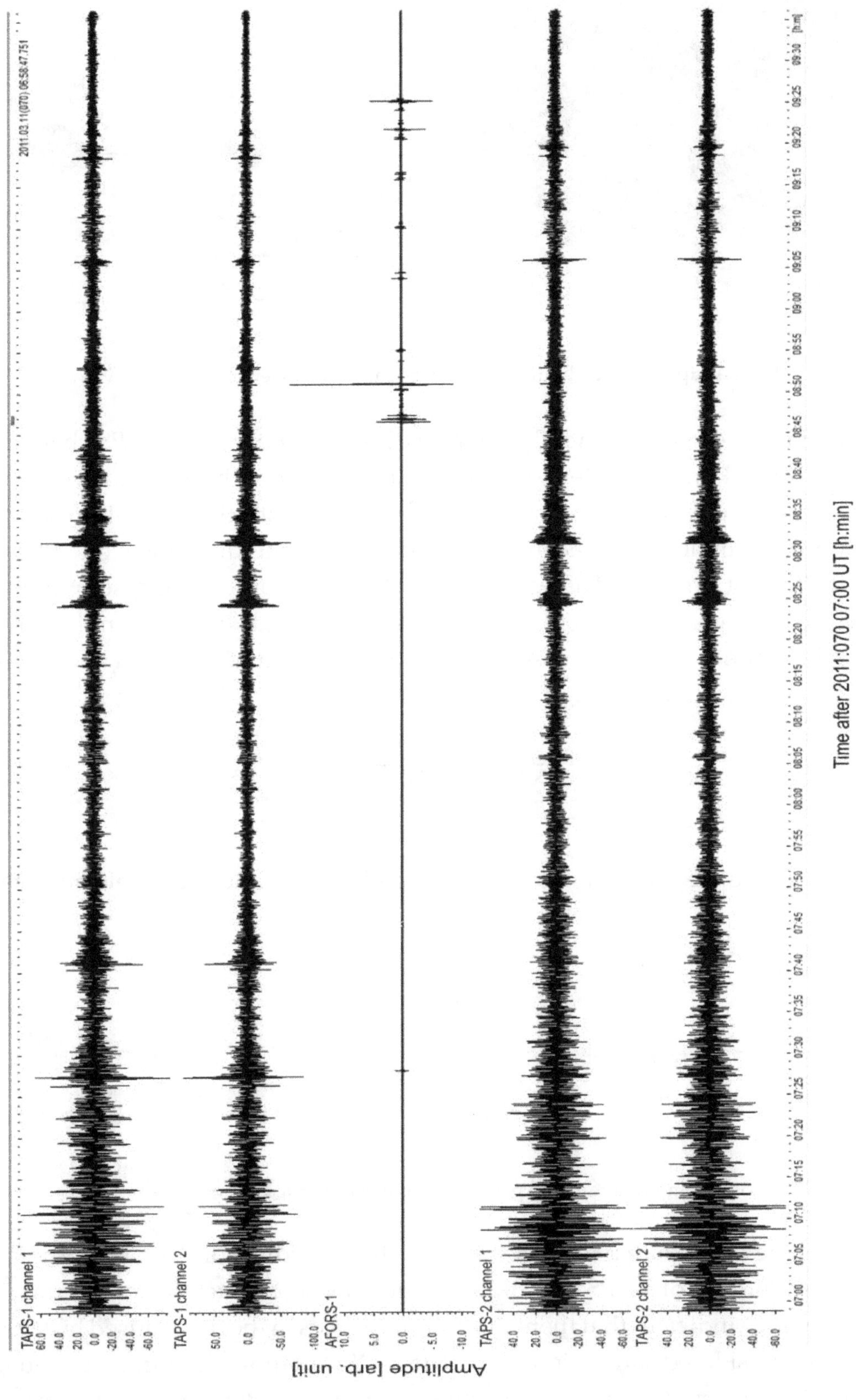

Figure 2. Plots of the seismic events recorded in the Książ Observatory, Poland on March 11th, 2011 starting from 06:58 UT, after the Honshu M=9.0 earthquake [30]

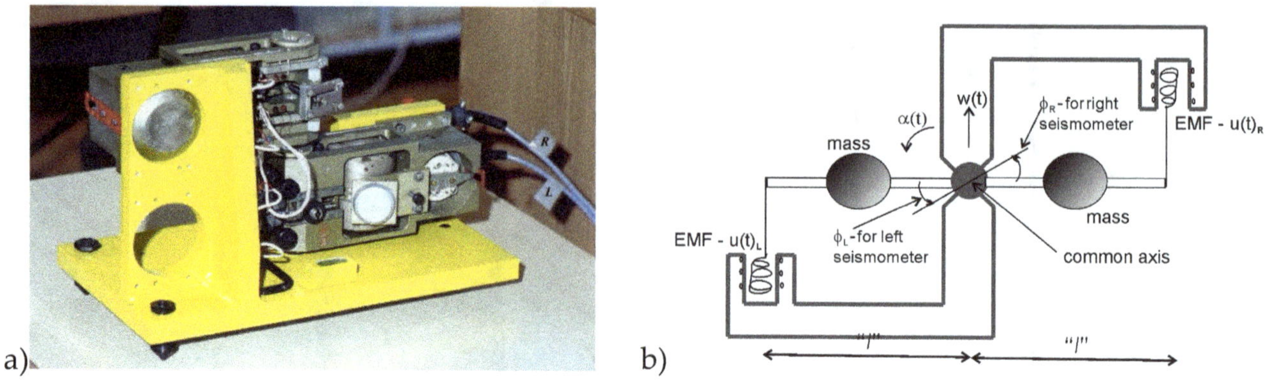

Figure 3. The TAPS – Twin Antiparallel Pendulum Seismometer: a) general view, b) schematic view

pendulum seismometers are, inevitably, slightly different, the special TAPS channels calibration algorithm is used. In this system, for the aim of comparing both sensors, it is possible to rotate the position of one seismometer in such a way that both the pendulum seismometers, suspended on the common axis become oriented in the same directions, one just above the other – this is the test position. The working and test position for the case of the horizontal seismometers are schematically shown in Figure 4. The records obtained in the test positions can differ mainly due to differences in their response characteristics, and to minimize these errors, the following left channel signal calibration procedure is usually applied:

$$u_L^{'} = u_L \sqrt{\sum u_R \cdot u_R / \sum u_L \cdot u_L}, \tag{4}$$

where: u_L and u_R are electromotive forces recorded by left and right seismometers in test position.

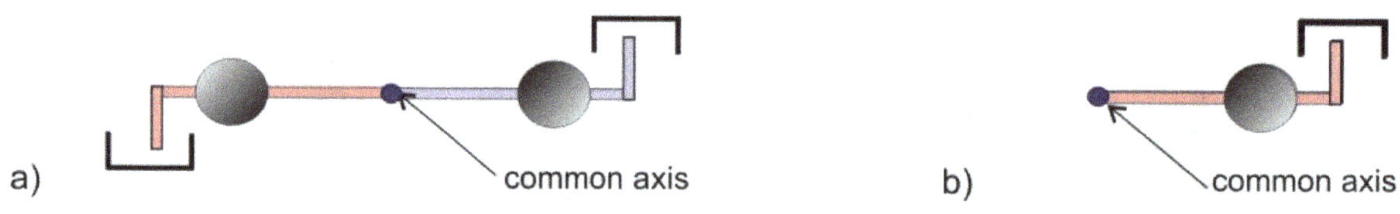

Figure 4. The TAPS: working a) and calibration b) positions of the pendulum seismometers

Nevertheless, there is some discrepancies due to the recording procedure in the case of the present of the difference of TAPS' pendulums attenuation characteristics. Figure 5 presents the simulation which indicates that there can be some errors in the data caused by the recording proceeding. The considered simulation was made for the attenuation difference equals $|\beta_L - \beta_R|$ =0.05 between left and right seismometer attenuation. It is easy to see that the major error of the signal is present when the simulated rotation is characterized by smaller value of amplitude compared to the simulated translation component. However, this is the region where the seismic-origin rotation is expected. For above reason the process of TASP calibration seems to be an essential complication of the system work's correctness. Moreover, the extremely high sensitivity to the translational motions of the seismometers (preferred for the component of displacement detection) taken into account in their construction can limit the accuracy of such devices, too.

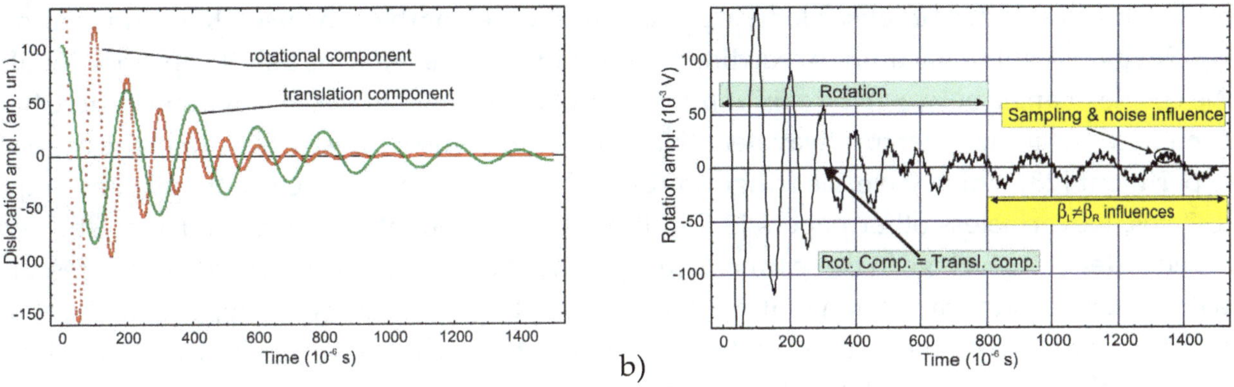

Figure 5. The simulated displacement and rotational components of the seismic event - a) and rotational signal detection by the TAPS - b) [22]

From above mentioned reasons, the additional numerical procedures for improving the TAPS performance may be applied, based on filtering in frequency [34] or in time [35] domains. The respective filters can then be applied to records in the normal working position to reduce the influences of non-equal operation of pendulum seismometers, presented above. However, these methods use test position of the TAPS, that generally changes the condition of the TAPS operation. For this reason, another procedure of the recorded data processing, based on smoothing by the spline functions has been also proposed [36]. It should be noticed, that the main disadvantage of all listed methods is that they operate on recorded data, which can limit TAPS usefulness for some applications. In the research presented in this paper, the mentioned methods of signal correction were not used.

2.2. Design of the Autonomous Fibre-Optic Rotational Seismograph

The AFORS-1, used in our research, is one of three such devices existing in Poland and manufactured on the base of fibre-optic gyroscope, all dedicated for direct measurement of rotational components existing in seismic events and having accuracy below $5 \cdot 10^{-9}$ rad/s for 1 Hz detection band.

The physical principle for these devices is the Sagnac effect [23] which is a result of difference between two beams propagating around closed optical path, in opposite direction. Figure 6a presents the basic principle of the Sagnac's experiment. The input light beam is splitted by a beam splitter into a beam circulating in the loop in a clockwise - cw direction (Figure 6a - beam T) and a beam circulating in the same loop in a counterclockwise - ccw direction (Figure 6a – beam R). One can observe the interference pattern, in the output light, caused by the interference phenomenon of the two waves. In the case of the present of rotation with an angular rate represented by vector $\vec{\Omega}$ rad/s then a fringe shift ΔZ is observed in the output of the interferometer. The fringe shift is given by the formula [23]:

$$\Delta Z = \vec{\Omega} \bullet \vec{S} / (\lambda_0 \cdot c), \tag{5}$$

in which $\vec{s}$ is a vector of sensor loop area, λ_0 is the wavelength of the used light in vacuum, c is velocity of light in vacuum. According to the above formula a fringe shift is proportional to the cosine of angle between the axis of rotation and the normal to the sensor plane. There have been used the indigo mercury wavelength as well as the sensor area S = 866cm^2 in the Sagnac's experiment [23] which gave the fringe shift equals 0.07 fringes for the rotational speed equal to 2 rps. Nevertheless other papers [37] indicate that the possible detectable fringe shift was of the order of 0.01 fringe in those time. Thus, perhaps Sagnac has been carried out the research with a maximal accuracy. Sagnac also established that the effect does not depend on the shape of surface area S or on the location of the centre of rotation, whereas future investigation shown that this effect it does not depend on the presence of a commoving refracting medium in the path of the beam [37].

Figure 6b presents the Sagnac interferometer in the optical fibre solution which uses optical waveguide of the long length L wound on sensor loop of the diameter D which was shown firstly in 1976 [38]. In this approach, a phase shift $\Delta\phi$ is produced between cw and ccw propagating light, given by:

$$\Delta\phi = \frac{2\pi \cdot L \cdot D}{\lambda_0 \cdot c}\Omega, \tag{6}$$

where Ω is the rotational component perpendicular to the sensor loop. It is clearly to indicate that the sensitivity can be change by physical dimension of the sensor loop as well as by the length of the applied waveguide. It should be noticed that application of three such systems, which loops are jointly perpendicular, provides data about space vector of the rotation. One can obtain the position change in space by integrating the data in time domain. The above procedure is used in the configuration of the fibre optical gyroscope - FOG which now, nearly 40-years from 1976, is the best recognized interferometric sensor performed in the fibre-optic technology.

However, for a desired rotation rate in the range of $10^{-6} - 10^{-9}$ rad/s, the Sagnac effect generates a very small phase shift, so it is needful to separate and protect this effect from other disturbances so that the Sagnac effect is the unique nonreciprocal effect in the device. For this reason all FOGs use, shown in Figure 7, the reciprocal configuration [39] which is also called minimum gyro-configuration [40]. This configuration guarantees an ideal equilibrium of two counter-propagating beams in the interferometer by obtaining true single mode operation at the common input-output port of the system. It is not disturbed even by non-single mode operation in the another part of the interferometer.

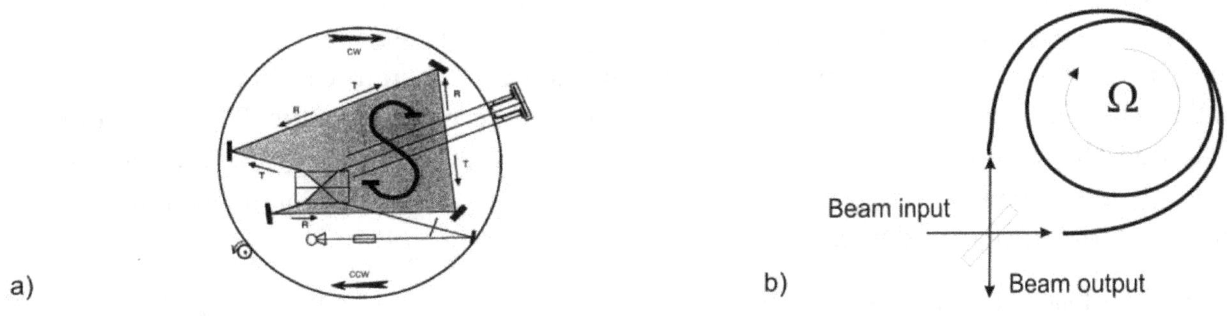

Figure 6. Schematic of the original Sagnac's experiment a) and its implementation in fibre optic technique b)

It is well known that each interferometric devices yield a cosine response. For above reason the detected signal practically do not change during the small changes of the rotation due to slow changes of the cosine function at the zero. In order to obtain higher sensitivity the operation point of the interferometer is shifted by applied additional phase shift modulation. The FOG utilizes the reciprocal phase modulator which is placed in the end of the sensor loop. It caused the modulation of the phase shift by propagation delay without any residual zero offset [41]. In this way one obtain the odd response instead of even one. An ultimate performance is, however, obtained only if the unbiased response is perfectly even and the biasing modulation has only odd frequencies. Therefore, the applied phase modulator is also a delay line filter operating at the eigen-frequency [42] – the delay in the loop is equal to a half of modulation period which suppresses the residual even harmonic signals. Nowadays the FOG utilizes broad-band light source for eliminating the Kerr effect which produces the phase shift in the optical fiber Sagnac interferometer [43]. Such a broadband source is also needed to remove coherence related with noise and drift due to backscattering and backreflection as well as lack of rejection of the polarizer [44–46]. Finally, for achieving the high scale factor linearization, FOG uses a digital phase step feedback [47] by the same reciprocal phase modulator as the biasing modulator and all-digital processing procedures where demodulation is carried by a digital subtracting and sampling of the modulated signal is obtained by using analogue-digital converter [48, 49].

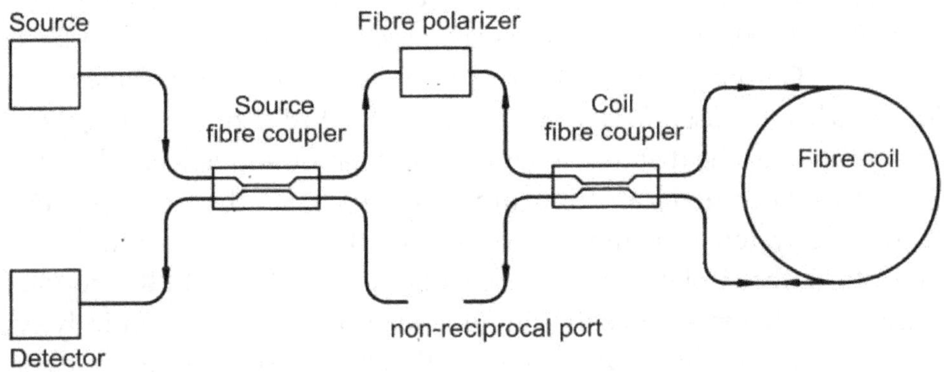

Figure 7. The minimum configuration of the FOG [39]

Currently, a digital processing of FOG systems is designed to record angular changes instead of rotation rates, thus, the optimization of such system to register the interesting phenomena from the rotational seismology point of view is problematic. Therefore it should be emphasized that the AFORS construction based on experiences according to the FOG development described above, but with system optimization for a direct measurement of the rotation rate only [22]. Such an approach gives a system which through a direct use of the Sagnac effect can limit drift influence on a device operation.

A detailed description of the AFORS system was published previously [29, 30, 50], hence here we summarized the above data regarding AFORS-1 construction, calibration and management. The second device - AFORS-2 is located in Warsaw (Poland) for initial works connected to the investigation of the irregular engineering construction torsional response and the interstory drift [51]. We anticipate that the new device, based on AFORS-1 and -2, AFORS-3 will be construed in the 2014 which gives us the opportunity to mount new innovative system instead of FORS-II assembled in seismological observatory Ojców, Poland [52].

The AFORS uses the minimum configuration of the FOG, however opposite of it, AFORS operates in open-loop architecture with digital data processing [53]. This technical solution is motivated by the fact that rotation events (Ω) are registered as sudden changes of rotational rate which amplitude is determined in a direct way from the Sagnac phase shift (Δφ) by following equation [37]:

$$\Omega = S_o \cdot \Delta\phi = \frac{\lambda \cdot c}{2\pi \cdot D \cdot L} \cdot \Delta\phi \tag{7}$$

where S_0 is the optical constant of interferometer which depends on the fundamental parameters of fibre coil.

Upper part of Figure 8 presents the block diagram of the AFORS-1 optical part configuration. The AFORS-1 construction, designed according to the minimum-gyro configuration, contains of the: SLED diode (ΔB=31.2 nm, λ_0 = 1305.7 nm, P_{out}= 9.43 mW; *Exalos*, Switzerland), isolator (α=0.34 dB, 39 dB isolation; *FCA*, Poland), depolarizator (DOP<5%, α=0.20 dB; *Phoenix Photonics*, UK), two mounted-in-line polarizators (ε=43 dB, α=0.45 dB each; *Phoenix Photonics*, UK), two X-couplers (α=0.20 dB; *Phoenix Photonics*, UK), sensor loop, phase modulator (*Piezomechanik GmbH*, Germany) and detector (S=0.9 A/W; *Optoway Technology Inc.*, Taiwan). The SLED diode has been chosen for two reasons. Firstly, SLED diode is the broad-band light source, which minimalizes the disadvantageous polarization as well as coherence effects [54, 55]. Additionally, this diode gives an opportunity to obtain a high optical power, which has a direct influence on the system sensitivity. In order to isolate the diode from the backscattering we applied the optical fibre isolator. To ensure truly depolarized light the depolarizer is used before the polarizers which guarantee the single mode operation in the entirely system as well as fulfil function of the filter. Two couplers ensure that both propagating waves have the same optical path. The detector's system consists of a PIN diode and a preamplifier. The sensor loop has been made by winding 15 000 m of the SMF-28e+ length on a special composite material which includes permalloy particles for shielding the system from external magnetic field. The double-quadrupole method of winding [56] was used in order to stabilize the work system during the temperature fluctuation. The technical optimization of AFORS-1 construction (optical fibre of 15 000 m with attenuation equals 0.451 dB/km in 0.63 m sensor loop) allows to obtain a theoretical sensitivity equal to $1.97 \cdot 10^{-9}$ rad/s/Hz$^{1/2}$ in quantum noise limitation.

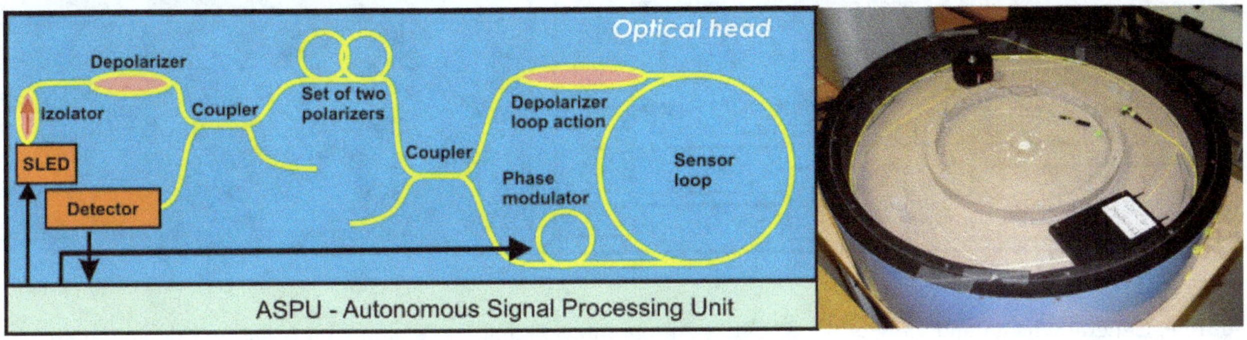

Figure 8. General schema and view of the AFORS optical head (generation of the Sagnac phase shift proportional to the measured rotation rate)

It should be emphasised that in the AFORS construction we have applied the special processing unit ASPU (Figure 9), which enables to obtain the detected rate of rotation in a direct way from the measured Sagnac phase shift (7). The ASPU detects the rotation rate (Ω) by selection and conversion of the first ($A_{1\omega}$) and second ($A_{2\omega}$) amplitude of the harmonic output signal [u(t)] using the following formula [50]:

$$\Omega = S_o \cdot \arctan[S_e \cdot u(t)] = S_o \cdot \arctan[S_e \cdot \frac{A_{1\omega}}{A_{2\omega}}], \tag{8}$$

in which S_e is the electrical constant connected to specification of the used instrumentations and received by calibrating the sensor, while S_o is the optical constant determined also during the calibration process which is described below. In order to eliminate a discontinuity the formula (8) uses the arc tangent which is extended to the four quarters ($-\pi$, π) after the Fourier transform. The ASPU utilizes the synchronous detection in a digital form. The 32 bit digital signal processor realizes all necessary processing processes including calculation the rate of rotation in a real time as well as data recording in the SD card. The electronic part sends also the stored data via fibre patch cord to the GSM/GPS modem which is connected FORS-Telemetric Server [30, 50]. The fundamental time constant equals 4.7104 ms and its value resulted from the used quartz oscillator. The multiplication (maximum multiplication is equal to 2^7) of the time constant allows to adjust the sampling time. This approach gives detection frequency bandpass from 0.83 Hz to 106.15 Hz. This detection band is required to detect the rotational seismic-origin events. Furthermore the electronic part is equipped with digital-to-analogue converter which enable to store the data in the analogue form by the standard recording system KST.

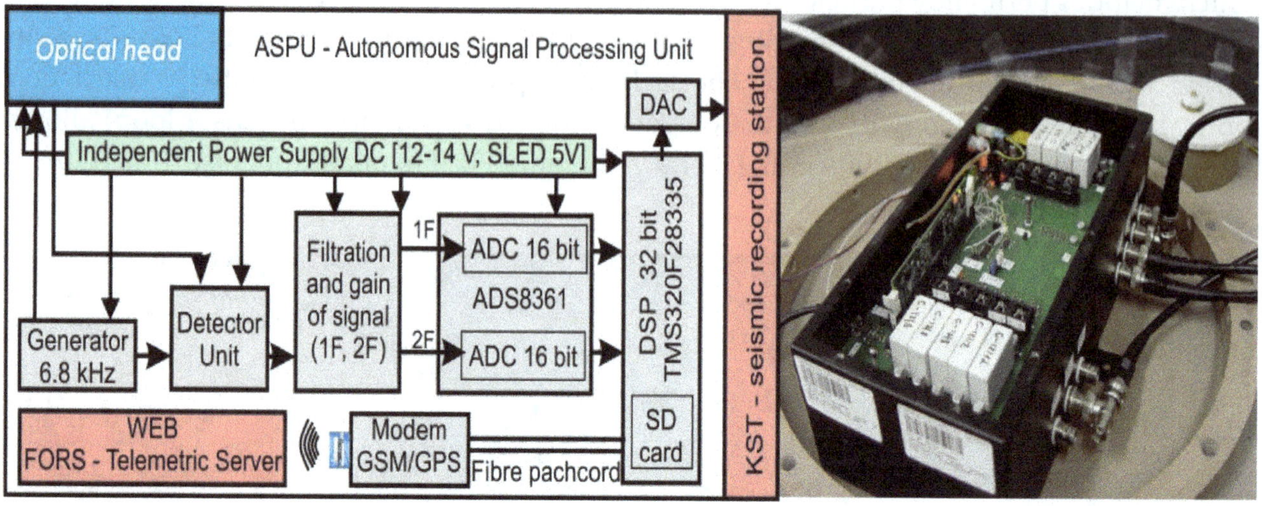

Figure 9. General schema and view of the AFORS's Autonomous Signal Processing Unit (rotation calculation and recording)

The calibration process was realized basing on the measurement of the defined slow rotation connected to the vector of Earth rotation in Warsaw, Poland (i.e. $\Omega_E = 9.18$ deg/h $\equiv 4.45 \bullet 10^{-5}$rad/s for $\phi = 52°20''$). AFORS was mounted vertically on a special rotation table (Figure 10a) and then rotated so that the sensor loop was directed to the North, South, East and West. The detected signal was equalled to zero for East-West (sensor loop collinear with the Earth rotation axis) while the signal was maximal, equal to $\pm 4.45 \bullet 10^{-5}$rad/s, for North-South (the sensor loop perpendicular to the vector component of Earth rotation). During the calibration procedure in the first step were determined the position for maximal signals, North and South, then, the position of the sensor for the West and East was defined as the midpoint between those two signals because of problems with determination this position basing on searching of signals equal zero. The maximal signals were obtained with accuracy equals 0.5 deg. In order to eliminate the drift phenomenon the 10 night hours averaging signal was applied. The above procedure allows for scaling the system and determining the constant operation parameters – optical S_o and electrical S_e. For AFORS-1 the following values of the above constants were obtained: $S_0 = 0.043$ s^{-1}, $S_e = 0.0144$. In order to experimental determination of the accuracy for particular device [50], which have been made according to AFORS production technology, we measured the noise level for each of them. Nevertheless the measurements were carried out at Military University of Technology, Warsaw, Poland. The place of the research could provide deviations due to urban noises. The received accuracy is equal to $5.07 \cdot 10^{-9}$ rad/s and $5.51 \cdot 10^{-8}$ rad/s, respectively, for the lower and higher working detection frequency band, as shown in Figure 10b. Additional Figure 10b shows also measured parameters for second system - AFORS-2 (theoretical sensitivity in quantum noise limitation equals $2.46 \cdot 10^{-9}$ rad/s/Hz$^{1/2}$, accuracy in detection band at level of $4.81 \cdot 10^{-9}$ rad/s and $6.11 \cdot 10^{-8}$ rad/s). It should be noted that the linear dependence of AFORSs accuracy on the detection frequency band is an advantage of this system.

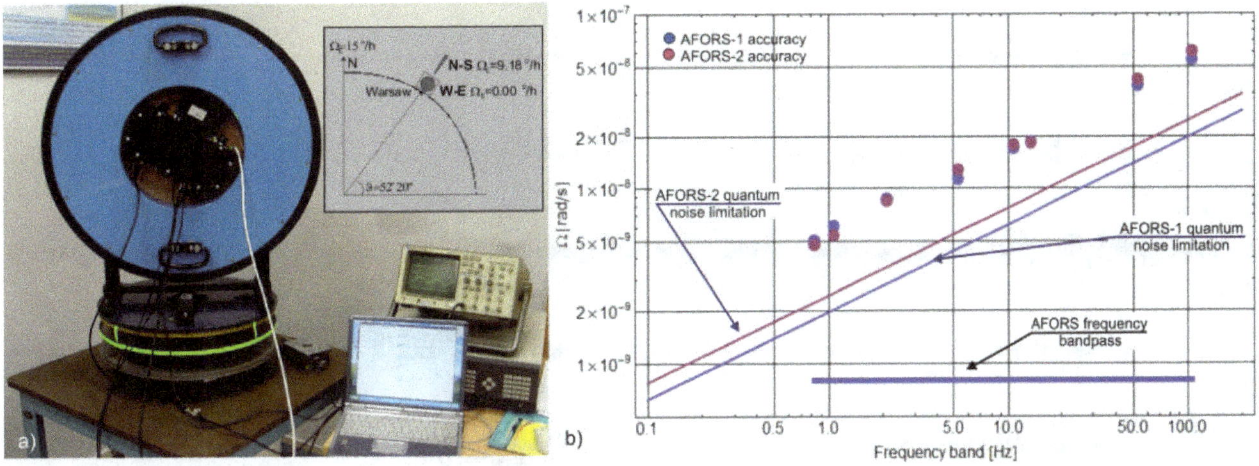

Figure 10. The calibration and investigation of AFORSs accuracies: a) general view of AFORS-1 during the calibration process with scheme showing its idea, b) the accuracy measured for the chosen detection frequency band for AFORS-1 and AFORS-2

3. Analysis of rotational components of the seismic fields caused by local events

In this section we present analysis of data obtained at Książ Observatory, reveal the rotational components presence in entire seismograms (from P-wave arrival onwards), for the cases of local seismic events, of the mining (Lubin on January 20th, 2011, two events starting at 04 h 59 min. 1 s UT - shown below as Figure 11) and tectonic (Jarocin on January 6th, 2012, the event starting at 15 h 38 min. 10 s UT - shown below as Figure 12) provenience.

The results were obtained directly from KST recording system and they include five plots: channels 1 and 2 for TAPS-1; channel 3 for AFORS-1; channels 4 and 5 for TAPS-2. The TAPSs' records show linear motions that appeared during the earthquakes, and rotational components are calculated as described in the section 2.1. The channel 3 for AFORS-1 shows rotational oscillations measured in a direct way.

As one can see, the both kinds of devices (AFORS-1 as well as set of micro-array of TAPSs) recorded the events in the same time, which can confirm some correlation between devices. However the following investigation needs an additional data proceedings. To limit noise influence on recorded signals the average procedure in the beginning has been applied. From above mentioned reason the recorded seismic events have been averaged in moving windows of 100 samples (which is equal to period of 1 second).

The results of above operation are presented in Figures 13a-13c respectively for above three events. Consistently, we present five plots for all of them. The first plot, named TAPS-1 channel 2, presents the velocity of linear ground motion in m/s registered by second seismometer in this device. Second diagram shows channel 1 of TAPS-2. It should be noticed that second channel for any TAPS, after change of sign, has very similar plot to the first channel, so they

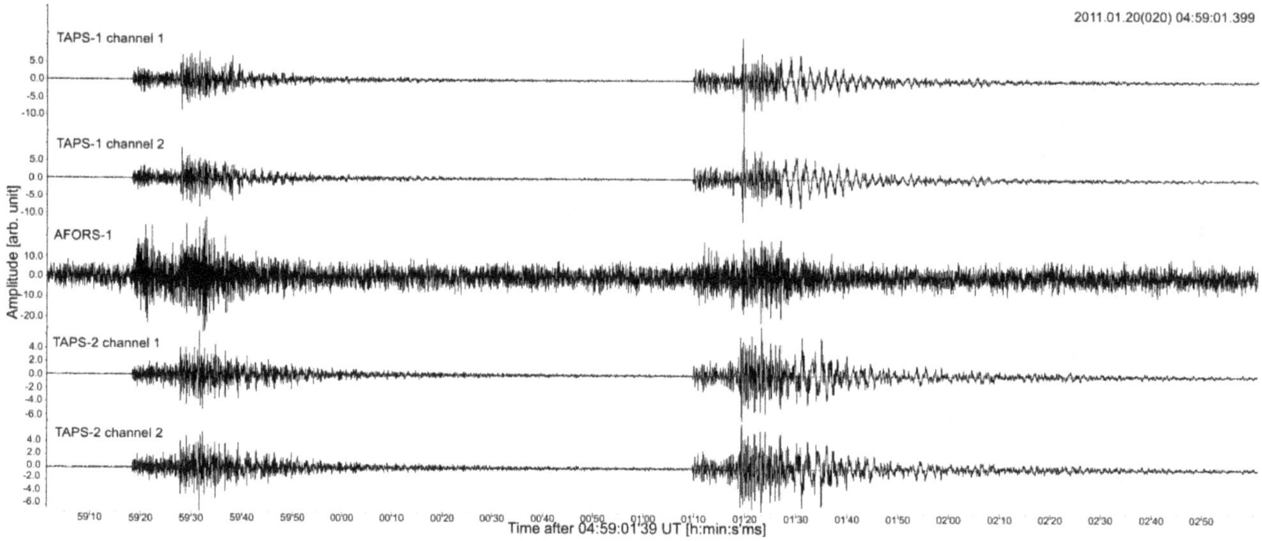

Figure 11. Plots of two seismic events with source in Lubin area: TAPS – ground motion velocity, AFORS – ground motion rotation rate

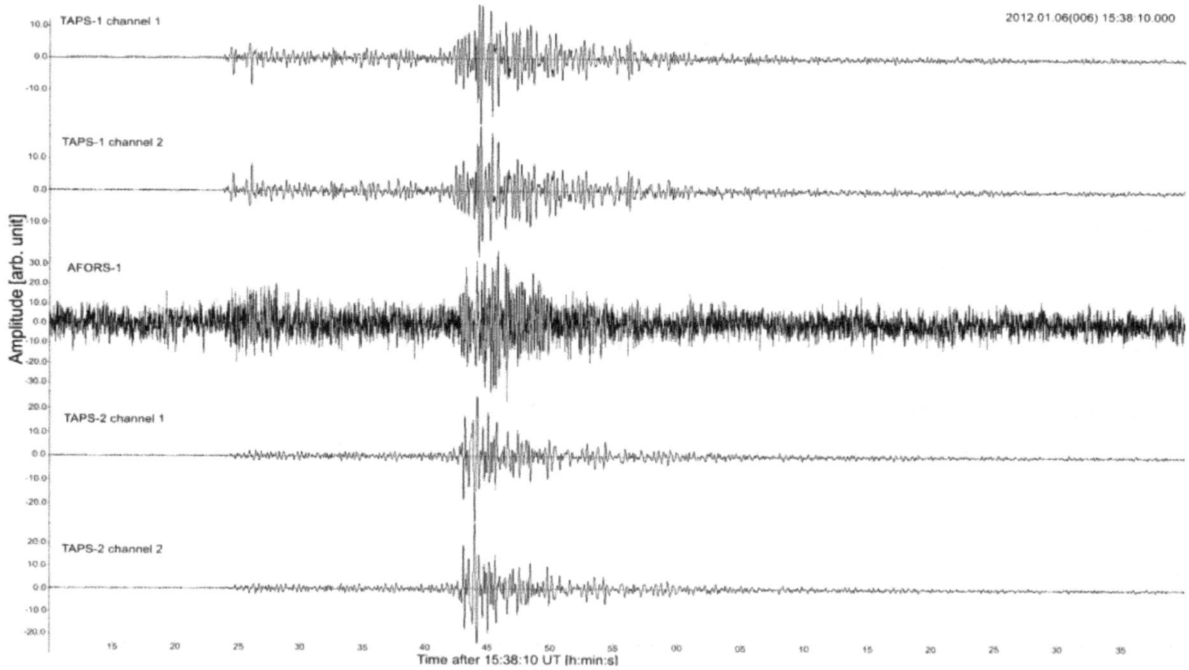

Figure 12. Plots of the seismic event with source in Jarocin area: TAPS – ground motion velocity, AFORS – ground motion rotation rate

are not presented both in our Figures. Third plot, named AFORS, presents the rotation velocity in rad/s, registered directly by AFORS-1. The result of measurement of the same component, but using the four simultaneous signals from micro-array of TAPSs (the procedure is described in section 2.1), is presented as plot four named Rotation from TAPS. Finally, last plot named Twist from TAPS, presents twist component obtained from the same micro-array in accord with, mentioned in the introduction, the Asymmetric Continuum Theory [14] – also in rad/s.

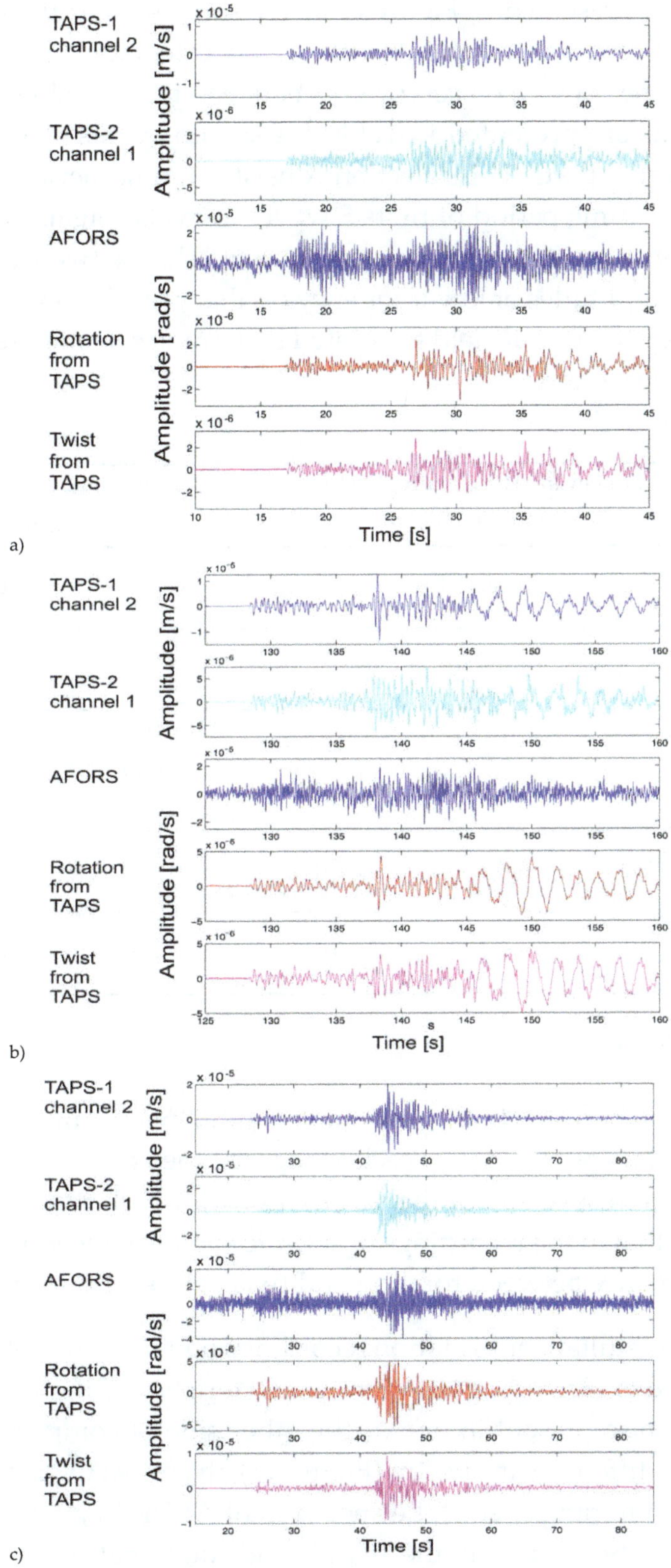

Figure 13. (a). Results for the first event in the Lubin area registered on January 20th, 2011; (b). Results for the second event in the Lubin area registered on January 20th, 2011; (c). Results for the Jarocin area earthquake registered on January 6th, 2012

Even though the results presented in Figures 13a-13c show similar time of occurrence of the rotational components recorded by the AFORS-1 and calculated from TAPSs, their shape differs, and inevitably the correlation coefficients between them are low (though greater than zero). These coefficients are presented in Table 1, for both events recorded on January 20th, 2011, and for their selected parts. This selection consists of: time-period when P waves arrive – 2 seconds in each case; time period of firsts S waves arrivals – again 2 seconds, and a time-period when great S-type oscillations dominate; here we choose twenty seconds in each case (such unusually high amplitudes of low frequency oscillations, dominating in the late stage of the tremor, characterize the seismic field generated by mining seismic events in the Lubin area and received at Książ observatory).

20.01.2011. Lubin	**First event 04 h 59 min., M=3.1 17 – 107.75 s**			**Second event 05 h 01 min., M=3.3 128.40 – 234.45 s**		
AFORS – Rot TAPS	0.092			0.063		
Twist TAPS – ROT TAPS	0.600			0.894		
Wave type time-period	P 17.00 – 19.00 s	S 26.00 – 28.00 s	great S 30.00 – 50.00 s	P 128.40 – 130.40 s	S 137.50 – 139.50 s	great S 146.00 – 166.00 s
AFORS – Rot TAPS	0.200	0.354	0.073	0.187	0.050	0.140
Twist TAPS – Rot TAPS	0.166	0.483	0.641	0.534	0.724	0.958

Table 1. The correlation coefficients

Additionally, correlation between both rotational components calculated from TAPSs recordings, that is – rotation and twist – was checked too, for the same chosen time-periods. This appeared generally high (especially in the second event), which confirms previous observations by KP. Teisseyre that in the recordings of seismic events made with a set of TAPSs, there usually occurs conformance between rotation and twist, either direct or reverse.

Dissimilarity between simultaneous rotation recordings obtained from AFORS and the micro-array of TAPSs may be explained by different characteristics of the used instruments or by certain errors or/and noise present in one or the other side of compared results, or in both. Here, there is dramatic difference in spectra (Figure 14), that of AFORS is always much longer. For the analyzed case of mining events, the spectrum of AFORS signal bears high amplitudes in the range of 2 – 8 Hz. Spectra of all the signals and rotational components obtained from TAPSs are short – they practically decrease to zero level at about 20 Hz – and bear sharp maximum at about 0.5 Hz, while the linear signals have also wide area of relatively high amplitudes in the range of 1 – 8 Hz. Moreover, the signals from AFORS bear high percentage of noise, probably of electronic provenience.

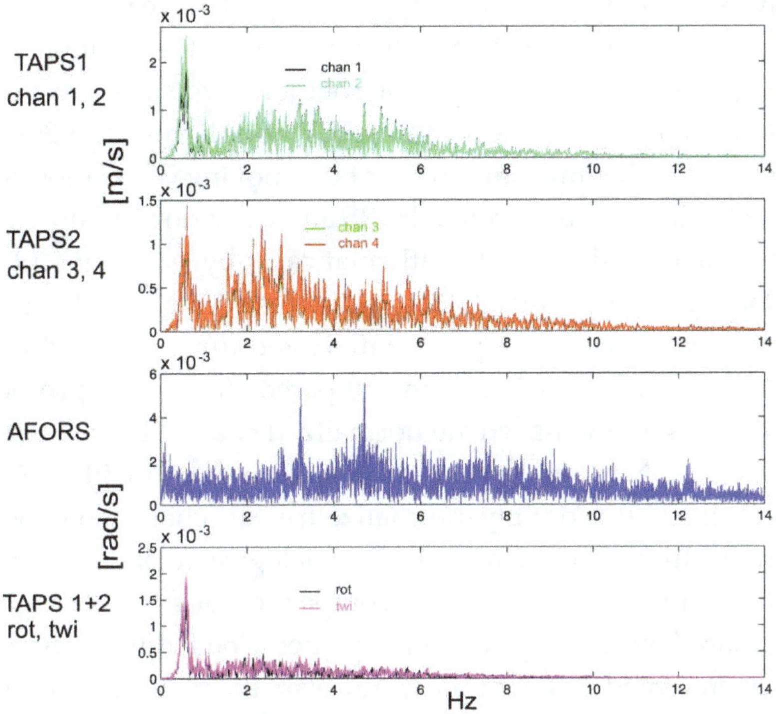

Figure 14. Spectra of linear and rotational signals, from ones of analysed seismic mining events, recorded on January 20th, 2011 at Książ observatory. The second one has the similar spectra

To find similarities in the obtained results in other way than just by sight, the following analysis has been applied. First – any of compared data chains was transformed into chain of absolute values (moduli). Then, the chains of moving averages of these moduli are created, with the window length of 100 samples, which is equivalent to 1 s. Here, absolute values of rotation (velocities) were compared – these from AFORS with those calculated from set of TAPSs. Further, analogical moving averages were investigated, but calculated from moduli of all four signals recorded by the TAPSs. If the rotation obtained from AFORS is symbolized with ω^o, rotation obtained from the TAPSs with ω, mean signal modulus from micro-array at the given sample with $\bar{u}$, and length of window used in moving average calculation with w, then formulae for described moving averages are as follows:

$$\bar{\omega}^0 = \frac{\sum_{i=1}^{i=w} |\omega_i^0|}{w}, \qquad \text{a}$$

$$\bar{\omega} = \frac{\sum_{i=1}^{i=w} |\omega_i|}{w}, \qquad \text{b} \qquad (9)$$

$$\bar{u} = \frac{\sum_{i=1}^{i=w} \left(\frac{|u_1| + |u_2| + |u_3| + |u_4|}{4} \right)_i}{w}. \qquad \text{c}$$

Results of this analysis are shown in Figures 15-17, which all have the same scheme. Diagrams a) show the plot of moving average of absolute values of the rotation rate obtained from AFORS; diagrams c) – analogical averages of rotation rate calculated from TAPSs' data; diagrams e) – analogical averages of the mean signals moduli from TAPSs – as in formula (9c). The blue plots presented in diagrams b) represent the moving correlation coefficient between moving averages of absolute rotation from AFORS and its analogue from TAPSs (compare the plots in diagrams a) and c) described mathematically by equations (9a) and (9b)). This coefficient is calculated for a window of 100 samples (1s). Additionally, in the same diagram we present reference thick lines – orange or mauve, and turquoise – which should facilitate the comparison of correlation coefficient with compared chains. The procedure for obtaining the upper reference line is following: create normalized chain (9b) to (9a) by comparing both maximum values, as (9b'). Next, make new chain as sum of (9a) with (9b'), and normalize it to 1. Thus we have obtained a doubly–normalized mean chain which joins shapes of the original two chains. The turquoise reference line is analogical to orange one, but multiplied by -1. Purpose of this line presence is to facilitate comparison of stages of high negative correlations between compared chains, again with the described doubly–normalized averages. Analogically, curves presented in diagrams d) and f) are the plots of the correlation coefficient in a moving window of 100 samples; in d) – between averages plotted in diagrams c) and e), and in f) – between averages plotted in e) and a). Reference lines are produced in analogical way as for diagrams b); the upper one is mauve in diagrams f). Figure 15 shows results of such analysis applied to recordings of first event from Lubin area; Figures 16 and 17 are made for second event from Lubin area and for Jarocin earthquake, in the same methodology.

Comparison of the moving correlation coefficients with the moving averages allows to find certain rule. In time-periods when main seismic phase arrive – as P and S (for local distances, it might be jointly for example Pg and Pb phases and analogically Sg and Sb in the S-type phases family), all investigated correlations between moving averages of the absolute signals are generally high. These are some of the time-periods in which blue line in the diagrams b), d) and f) is near 1, and the upper reference line rises. More interestingly, these time-periods starts slightly before any noticeable rise in initial moving averages, despite facts that all the moving windows had the same length. This phenomenon may have two causes: certain common order in concerned signals starts just before noticeable arrival of seismic waves, or/and this is a mere effect of filtration (which used in every contemporary seismic recording system).

The chosen time-periods of high correlations and rise of the moving averages are as follows.

For the first seismic event from Lubin - two time periods: from 16.2 to 18.5 s, this include arrival of P waves, and from 25.8 to 27 s, which include S waves arrival (see Table 1). For the second seismic event from Lubin – only one time period, from 127.9 to 129.5 s, which include arrival of P waves. For the Jarocin earthquake – one non-continuous time period, for diagrams b) and f): 21.7 – 24.1 s and for diagram d) : 20.9 – 24.1 s. This time-period, in all three diagrams, starts before arrival of P waves and comprises arrival of this seismic phase.

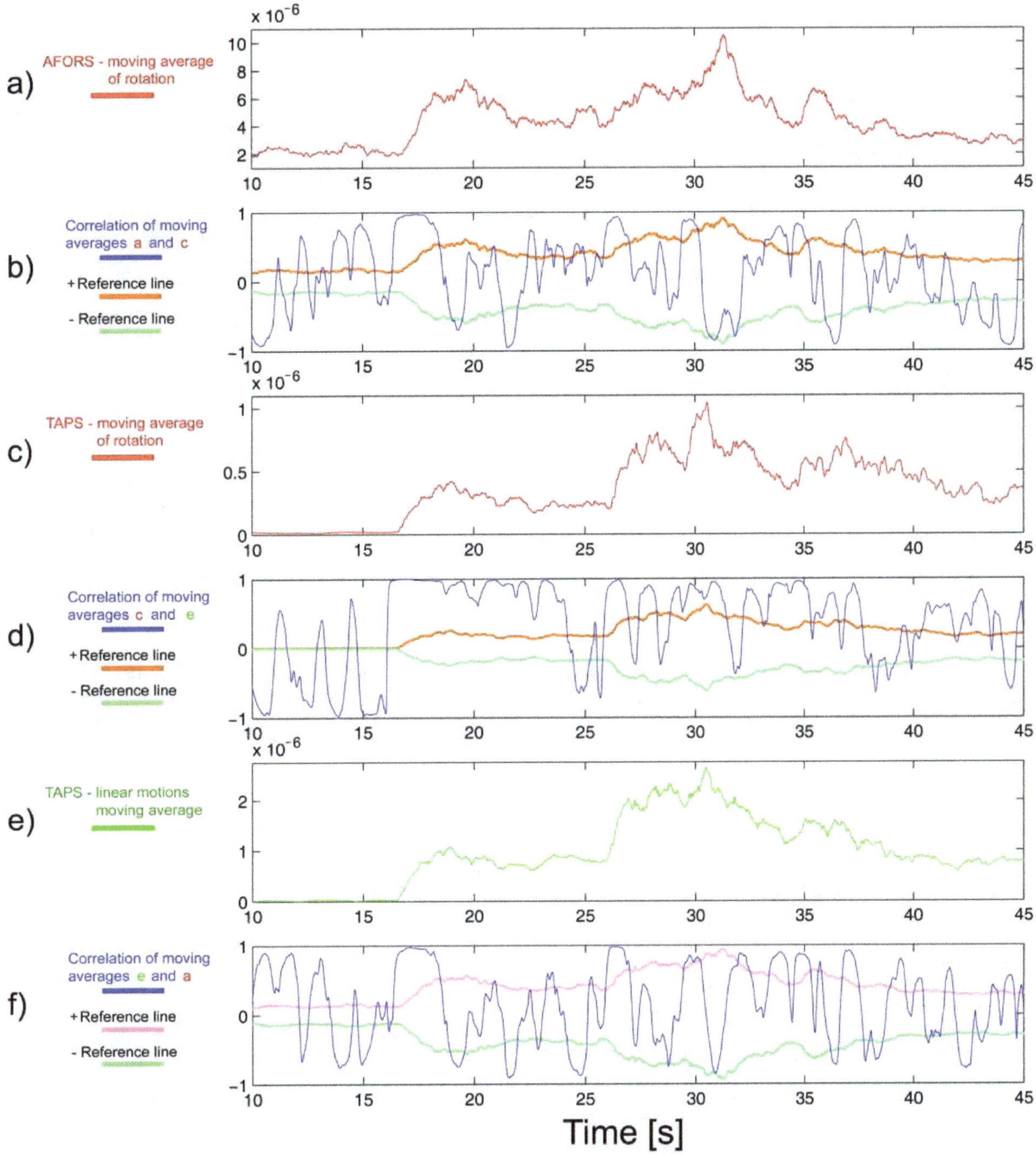

Figure 15. Relations of rotations and linear signals at the first event in Lubin area registered on January 20th, 2011. Diagrams: a) moving average of absolute values of AFORS – signal; b) correlation coefficient between (a) and (c) in moving window – blue line, orange reference line – doubly normalized means of (a) and (c), turquoise – the same as orange but with reversed sign; c) moving average of absolute values of rotations obtained from TAPSs; d) correlation coefficient between (c) and (e) in moving window, orange reference line – doubly normalized means of (c) and (e), turquoise – the same as orange but with reversed sign; e) moving average of absolute values of four channels of TAPSs; f) correlation coefficient between (e) and (a) in moving window, mauve reference line – doubly normalized means of (e) and (a), turquoise – the same as mauve but with reversed sign

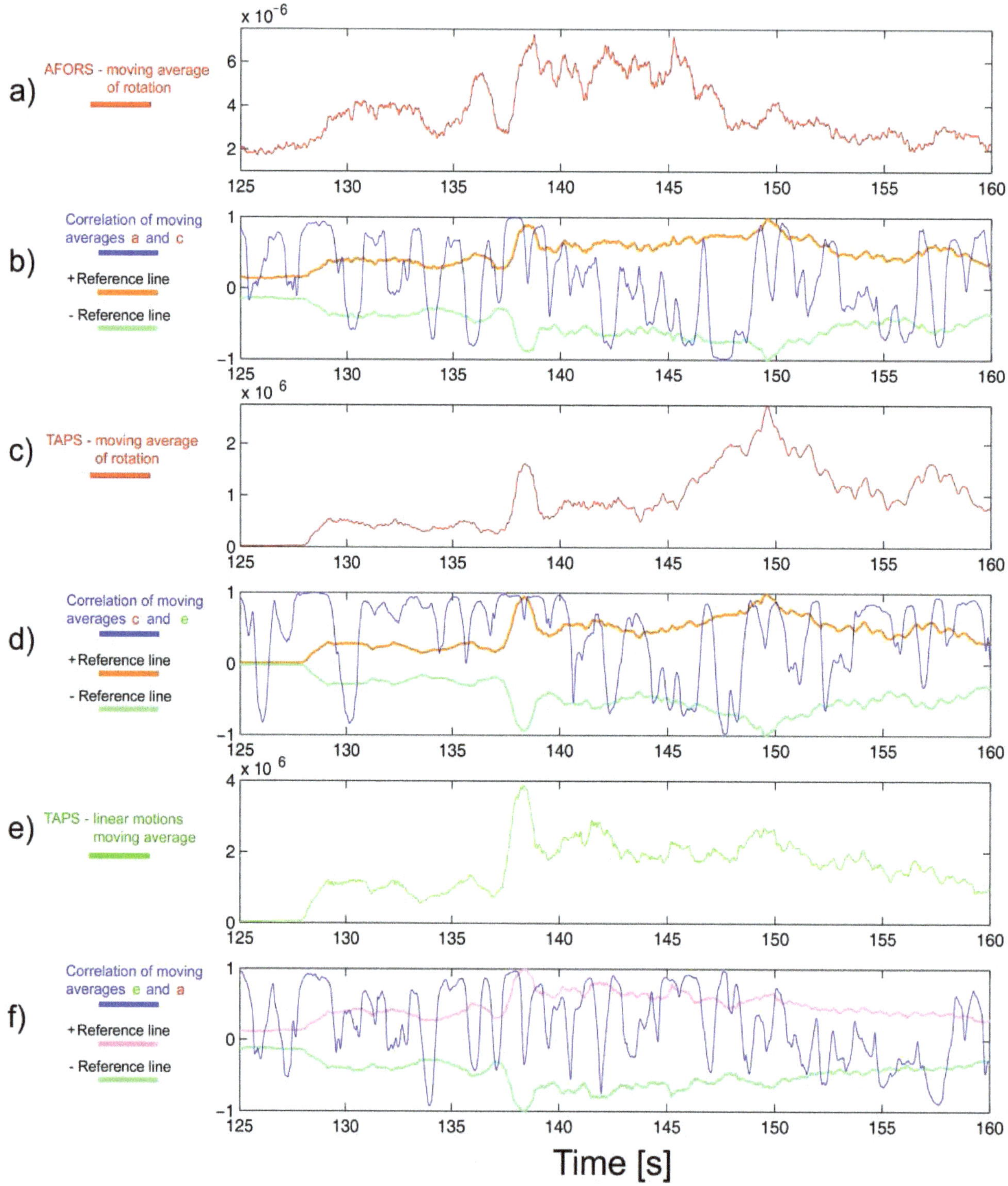

Figure 16. Relations of rotations and linear signals at the second event in Lubin area registered on January 20th, 2011. Diagrams descriptions as in Fig. 15

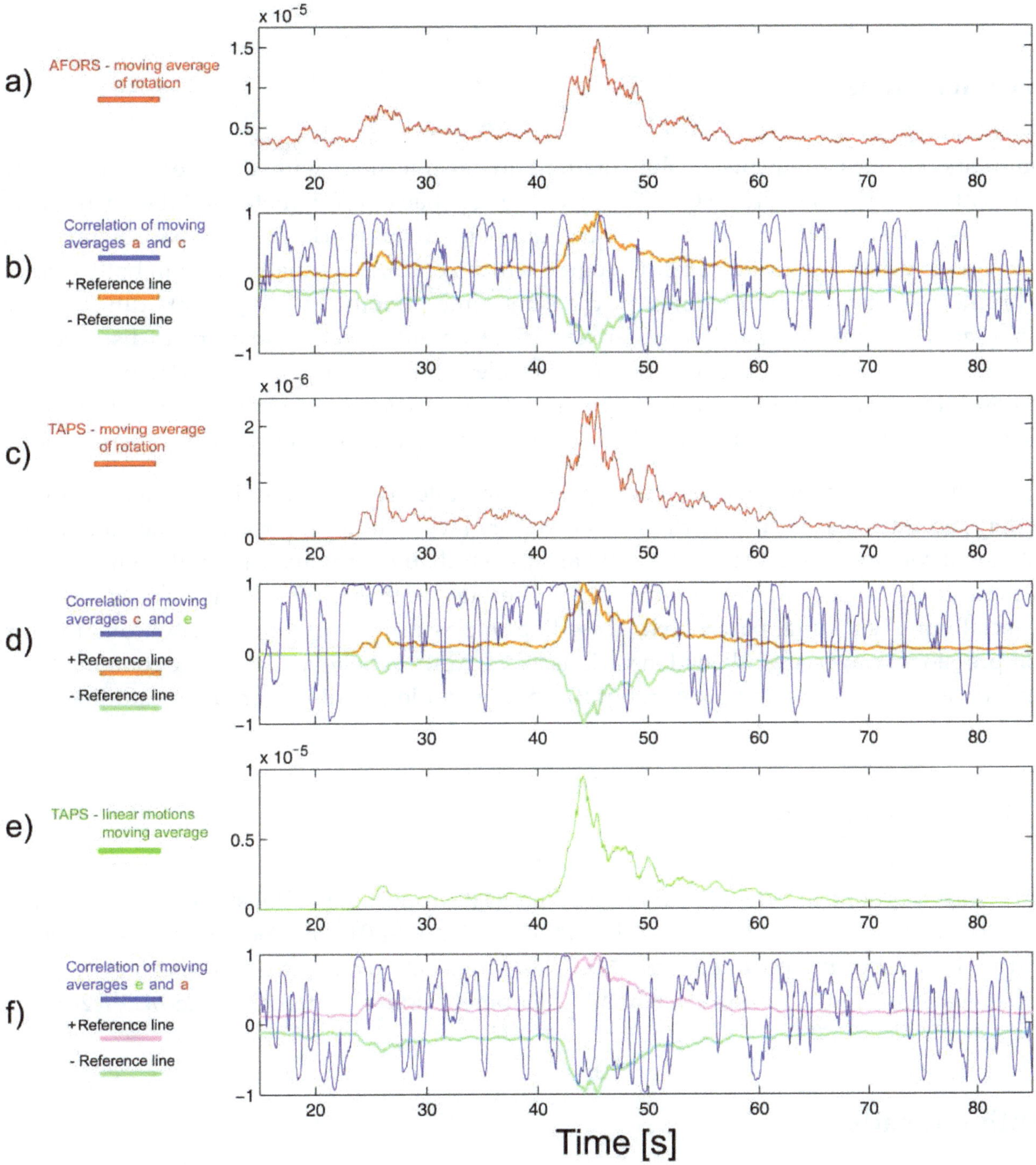

Figure 17. Relations of rotations and linear signals at the event in Jarocin area registered on January 6th, 2012. Diagrams descriptions as in Fig. 15

We did not find yet any other rules, especially – we did not find any relation between episodes of negative correlations (near -1) and the initial moving averages. We suppose that such comparison method may be useful in analysis of various chains, and especially their moving averages, as in this work. Crucial point is that not the original signals, but chains of their absolute values are compared.

4. Conclusions

Simultaneous measurements of the rotations in seismic field with the use of completely different instruments – here AFORS-1 which is the Sagnac interferometer and the micro-array of TAPSs allow for comparison of the used equipment. In this work, such comparison revealed that signals differ significantly, to the degree which complicates analysis. From both kind of instruments, rotations are obtained in the same time-periods, but their plots differ. These differences are attributed partly to difference in instruments spectra and partly to disturbances in the signals, of technical provenience. Nevertheless, an analysis using the moving averages of absolute signals values, and consequently also coefficients of correlation between these averages confirmed common roots of these recorded signals, despite all their imperfections.

Research on seismic rotational effects, especially in buildings and other large constructions is widely recognized as very important and therefore these studies flourish. On the other hand, rotational components in the seismic field are also studied in various ways, but even existence of these components still evoke controversy. Presence of these components in seismograms, especially in their initial part which, according to classical theory of elasticity, contain only compressional waves, is explained in various ways. Authors believe that no one contemporary explanation is complete and proven, but this not preclude usefulness of further studies.

Acknowledgements

This work was done in 2013-2014 under the financial support of the Polish Ministry of Science and Higher Education under Key Project POIG.01.03.01-14-016/08 "New photonic materials and their advanced application", the MUT statutory activity PBS-850 and partially the the Polish National Centre for Research and Development under contract No PBS1/B3/7/2012.

Author details

Anna Kurzych[1], Krzysztof P. Teisseyre[2], Zbigniew Krajewski[1] and Leszek R. Jaroszewicz[1*]

*Address all correspondence to: jarosz@wat.edu.pl

1 Institute of Applied Physics, Military University of Technology, Warsaw, Poland

2 Institute of Geophysics, Polish Academy of Sciences, Warsaw, Poland

References

[1] Hobbs WH. Earthquakes. An Introduction to Seismic Geology. New York: Appleton and Co.; 1907.

[2] Davison Ch. The Founders of Seismology. Cambridge: Cambridge University Press; 1927.

[3] Ferrari G. Note on the Historical Rotation Seismographs. In: Teisseyre R., Takedo M., Majewski E. (eds) Earthquake Source Asymmetry, Structural Media and Rotation Effects. Berlin-Heidelberg-New York: Springer; 2006. p367-376.

[4] Kozak JT. Development of Earthquake Rotation Effect Study. In: Teisseyre R., Takedo M., Majewski E. (eds) Earthquake Source Asymmetry, Structural Media and Rotation Effects. Berlin-Heidelberg-New York: Springer; 2006. p3-10.

[5] Imamura A. Theoretical and Applied Seismology. Tokyo: Maruzen Co.;1937.

[6] Gutenberg B. Grundlagen der Erdbebrnkunde. Frankfurt: Univ. Frankfurt a/M; 1927.

[7] Teisseyre R. Earthquake Processes in a Micromorphic Continuum. Pure Applied Geophysics 1973; 102(1) 15-28.

[8] Teisseyre R. Symmetric Micromorphic Tin Space Continuum: Wave Propagation, Point Source Solutions and Some Applications to Earthquake Processes. In: Thoft-Christensen P. (ed.) Continuum Mechanics Aspects of Geodynamics and Rock Fracture Mechanics. Dodrecht-Boston: D. Reidel Publ. Comp.; 1974. p201-244.

[9] Eringen AC. Mirocontinuum Field Theories. Vol. 1 Foundations and Solids. New York: Springer-Verlag; 1999.

[10] Teisseyre R, Boratyński W. Continuum with Self-Rotation Fields: Evolution of Defect Fields and Equations of Motion. Acta Geophysica 2002; 50(3) 223-229.

[11] Teisseyre R. Asymmetric Continuum Mechanics: Deviations from Elasticity and Symmetry. Acta Geophysica 2005; 53(2) 115-126.

[12] Teisseyre R, Białecki M, Górski M. Degenerated Mechanics in a Homogeneous Continuum: Potentials for Spin and Twist. Acta Geophysica 2005; 53(3) 219-223.

[13] Teisseyre R, Górski M. Transport in Fracture Processes: Fragmentation of Defect Fields and Equations of Motion. Acta Geophysica 2009; 57(5) 583-599.

[14] Teisseyre R, Suchcicki J, Teisseyre KP, Wiszniowski J, Palangio P. Seismic Rotation Waves: Basic Elements of Theory and Recording. Annals of Geophysics 2003; 46(4) 671-685.

[15] Teisseyre R., Kozák JT. Sources of Rotation and Twist Motions. In: Teisseyre R, Takedo M., Majewski E. (eds) Earthquake Source Asymmetry, Structural Media and Rotation Effects. Berlin-Heidelberg-New York: Springer; 2006. p11-23.

[16] Teisseyre R, Takeo M, Majewski E. Earthquake Source Asymmetry, Structural Media and Rotation Effects. Berlin-Heidelberg-New York: Springer; 2006.

[17] Lee WHK, Cęlebi M, Todorovska MI, Igel H. (guest eds) Rotational Seismology and Engineering Applications. California: Bulletin of the Seismological Society of America 99(2B); 2009.

[18] Cowsik R, Madziwa-Nussinov T, Wagoner K, Wiens D, Wysession M. Performance Characteristics of a Rotational Seismometer for Near-Field and Engineering Applications. Bulletin of the Seismological Society of America 2009; 99(2B) 1181-1189.

[19] Droste Z, Teisseyre R. Rotational and Displacemental Components of Ground Motion as Deduced from Data of the Azimuth System of Seismograph. Publications od Institute of Geophysics Polish Academy of Science 1976; 97 157-167.

[20] Riedesel MA, Moore RD, Orcutt JA. Limits of Sensitivity of Inertial Seismometers and Velocity Transducer and Electronic Amplifiers. Bulletin of the Seismological Society of America 1990; 80 1725-1752.

[21] Teisseyre R, Nagahama H. Micro-Inertia Continuum: Rotations and Semi-Waves. Acta Geophysica Polonica 1999; 47 259–272.

[22] Jaroszewicz LR, Krajewski Z, Solarz L, Marc P, Kostrzyński T. A New Area of the Fiber-Optic Sagnac Interferometer Application. In: Proceedings of the 2003 SBMO/IEEE MTT-S International Conference on Microwave and Optoelectronics, 20–23 September 2003, Rio de Janeiro, Brazil, 2003.

[23] Sagnac G. L'éther lumineux démontré par l'effet du vent relative d'éther dans un interférométre en rotation uniforme. [in French] Comptes rendus de l'Académie des Sciences 1913; 95 708–710.

[24] Schreiber U, Schneider M, Rowe CH, Stedmanand GE, Schluter W. Aspects of Ring Lasers as Local Earth Rotation Sensors. Surveys in Geophysics 2001; 22 603–611.

[25] Jaroszewicz LR., Krajewski Z., Solarz L. Absolute Rotation Measurement Based on the Sagnac Effect. In: Teisseyre R., Takeo M., Majewski E. (eds) Earthquake Source Asymmetry, Structural Media and Rotation Effects. Berlin-Heidelberg-New York: Springer; 2006. p413–438.

[26] Takeo M, Ueda H, Matzuzawa T. Development of a High-Gain Rotational-Motion Seismograph. Grant 11354004. Earthquake Research Institute University of Tokyo; 2002. p5–29.

[27] Nigbor RL, Evans RJ, Hutt C. Laboratory and Field Testing of Commercial Rotational Seismometers. Bulletin of the Seismological Society of America 2009; 99, 1215–1227.

[28] Schreiber KU, Stedman GE, Igel H, Flaws A. Ring Laser Gyroscopes as Rotation Sensors for Seismic Wave Studies. In: Teisseyre R., Takeo M., Majewski E. (eds) Earth-

quake Source Asymmetry, Structural Media and Rotation Effects. Berlin-Heidelberg-New York: Springer; 2006. p377-390

[29] Kurzych A, Jaroszewicz LR, Krajewski Z, Teisseyre KP, Kowalski JK. Fibre Optic System for Monitoring Rotational Seismic Phenomena. Sensors 2014;14 5459-5469.

[30] Jaroszewicz LR, Krajewski Z, Teisseyre KP. Usefulness of AFORS - Autonomous Fibre-Optic Rotational Seismograph for Investigation of Rotational Phenomena. Journal of Seismology 2012; 16(4) 573-586.

[31] Teisseyre KP. Mining Tremors Registered at Ojców and Książ Observatories: Rotational Field Components, Publications of Institute Geophysics Polish Academy of Sciences 2006; M-29(395) 77-92.

[32] Teisseyre KP. Analysis of a Group of Seismic Events Using Rotational Components. Acta Geophysica 2007; 55(4) 535-553.

[33] Moriya T, Teisseyre R. Discussion on the Recording of Seismic Rotation Waves. Acta Geophysica Polonica 1999; 47 351-362.

[34] Teisseyre R, Suchcicki J, Teisseyre KP. Recording the Seismic Rotation Waves: Reliability Analysis. Acta Geophysica Polonica 2002; 51 37-50.

[35] Nowożyński K. Teisseyre KP. Time-Domain Filtering of Seismic Rotation Waves. Acta Geophysica Polonica 2003; 51 51-61.

[36] Solarz L, Krajewski Z, Jaroszewicz LR. Analysis of Seismic Rotation Detected by Two Antiparallel Seismometers: Spine Function Approximation of Rotation and Displacement Velocities. Acta Geophysica Polonica 2004; 52 198-217.

[37] Post EJ. Sagnac Effect. Review of Modern Physics 1967; 39(2) 475-493.

[38] Vali V, Shorthill RW. Fiber Ring Interferometer. Applied Optics 1976; 15(5) 1099-1100.

[39] Ulrich R. Fiber-Optic Rotation Sensing with Low Drift. Optics Letters 1980; 5(5) 173-175.

[40] Arditty H, Lefèvre HC. (1981). Sagnac Effect in a Fiber Gyroscope, Optics Letters 1981; 6(8) 401-403.

[41] Martin JM, Winkler JT. Fiber-Optic Laser Gyro Signal Detection and Processing Technique. SPIE Proceedings 1978; 139 98-102.

[42] Bergh RA, Lefèvre HC, Shaw HJ. All-Single-Mode Fiber-Optic Gyroscope with Long-Term Stability. Optics Letters 1981; 6(10) 502-504.

[43] Ezekiel S, Davis JL, Hellwarth RW. Intensity Dependent Nonreciprocal Phase Shift in a Fiber-Optic Gyroscope. Springer Series in Optical Sciences 1982; 32 332-336.

[44] Fredricks RJ, Ulrich R. Phase-Error Bounds of Fibre Gyro with Imperfect Polariser/ Depolarizer. Electronic Letters 1984; 20(8) 330-332.

[45] Lefèvre HC, Bettini JP, Vatoux S, Papuchon M. Progress in Optical Fiber Gyroscopes Using Integrated Optics. Proceedings of AGARD-NATO 1985; CPP-383 9A1-9A13.

[46] Burns WK. Phase-Error Bounds of Fiber Gyro With Polarization-Holding Fiber. Journal of Lightwave Technology 1986; LT4(1) 8-14.

[47] Lefèvre HC, Graindorge Ph, Arditty HJ, Vatoux S, Papuchon M. Double Closed-Loop Hybrid Fiber Gyroscope Using Digital Phase Ramp. Proceeding of OFS-3 1985, San Diego, OSA/IEEE, Postdeadline Paper 7

[48] Auch W. The Fiber-Optic Gyro – a Device for Laboratory Use Only?, SPIE Proceedings 1986; 719, 28-34.

[49] Arditty HJ, Graindorge Ph, Lefèvre HC, Martin Ph, Morisse J, Simonpiétri P. Fiber-Optic Gyroscope with All-Digital Processing. Proceedings of OFS- 6, Paris, Springer-Verlag Proceedings in Physics 1989; 44 131-136.

[50] Jaroszewicz LR, Krajewski Z, Kowalski H, Mazur G, Zinówko P, Kowalski JK. AFORS Autonomous Fibre-Optic Rotational Seismograph: Design and Application. Acta Geophysisica 2011; 59, 578–596.

[51] Jaroszewicz L.R., Krajewski Z., Teisseyre, K. P. The Possibility of a Continuous Monitoring of the Horizontal Buildings' Rotation by the Autonomous Fibre-Optic Rotational Seismograph AFORS Type. In: Lavan O., De Stefano M. (eds) Seismic Behaviour and Design of Irregular and Complex Civil Structures. Berlin-Heidelberg: Springer-Verlag; 2013. p339-351.

[52] Jaroszewicz LR, Krajewski Z. Application of the Fibre-Optic Rotational Seismometer in Investigation of the Seismic Rotational Waves. Opto-Electronics Review 2008; 16(3) 314-320.

[53] Böhm K, Marten P, Standigel L, Weidel E. Fiber-Optic Gyro with Digital Data Processing. 2nd International Conference on Optical Fiber Sensors, Stuttgart, Germany, 5–7 September 1984; 251–258.

[54] Krajewski Z. Fiber-Optic Sagnac Interferometer as System for Rotational Phenomena Investigation Connected with Seismic Events. (In Polish) PhD thesis, Military University of Technology, Warsaw; 2005.

[55] Krajewski Z, Jaroszewicz LR, Solarz L. Optimization of Fiber-Optic Sagnac Interferometer for Detection of Rotational Seismic Events. In Proceedings of SPIE 5952, Optical Fibers: Applications, Warsaw, Poland, 28 August–2 September 2005; 240–248.

[56] Dai X, Zhao X, Cai B, Yang G, Zhou K, Liu C. Quantitative Analysis of the Shupe Reduction in a Fiber Optic Sagnac Interferometer. Optical Engineering 2002; 41 1155–1156.

Permissions

All chapters in this book were first published in EEFESOSDES, by InTech Open; hereby published with permission under the Creative Commons Attribution License or equivalent. Every chapter published in this book has been scrutinized by our experts. Their significance has been extensively debated. The topics covered herein carry significant findings which will fuel the growth of the discipline. They may even be implemented as practical applications or may be referred to as a beginning point for another development.

The contributors of this book come from diverse backgrounds, making this book a truly international effort. This book will bring forth new frontiers with its revolutionizing research information and detailed analysis of the nascent developments around the world.

We would like to thank all the contributing authors for lending their expertise to make the book truly unique. They have played a crucial role in the development of this book. Without their invaluable contributions this book wouldn't have been possible. They have made vital efforts to compile up to date information on the varied aspects of this subject to make this book a valuable addition to the collection of many professionals and students.

This book was conceptualized with the vision of imparting up-to-date information and advanced data in this field. To ensure the same, a matchless editorial board was set up. Every individual on the board went through rigorous rounds of assessment to prove their worth. After which they invested a large part of their time researching and compiling the most relevant data for our readers.

The editorial board has been involved in producing this book since its inception. They have spent rigorous hours researching and exploring the diverse topics which have resulted in the successful publishing of this book. They have passed on their knowledge of decades through this book. To expedite this challenging task, the publisher supported the team at every step. A small team of assistant editors was also appointed to further simplify the editing procedure and attain best results for the readers.

Apart from the editorial board, the designing team has also invested a significant amount of their time in understanding the subject and creating the most relevant covers. They scrutinized every image to scout for the most suitable representation of the subject and create an appropriate cover for the book.

The publishing team has been an ardent support to the editorial, designing and production team. Their endless efforts to recruit the best for this project, has resulted in the accomplishment of this book. They are a veteran in the field of academics and their pool of knowledge is as vast as their experience in printing. Their expertise and guidance has proved useful at every step. Their uncompromising quality standards have made this book an exceptional effort. Their encouragement from time to time has been an inspiration for everyone.

The publisher and the editorial board hope that this book will prove to be a valuable piece of knowledge for researchers, students, practitioners and scholars across the globe.

List of Contributors

Constantine A. Stamatopoulos
Hellenic Open University, Athens, Greece
Stamatopoulos and Associates Co, Greece

Yan-Yan Huang
Automation School, Nanjing University of Science and Technology, Nanjing, Jiangsu, China

Seigo Sakai
Yokohama National University, Japan

Ming-Yi Liu and Pao-Hsii Wang
Department of Civil Engineering, Chung Yuan Christian University, Jhongli City, Taoyuan County, Taiwan

Ming-Ching T. Kuo
Department of Mineral and Petroleum Engineering, National Cheng Kung University, Tainan, Taiwan

Yingbin Zhang
Department of Geotechnical Engineering, School of Civil Engineering, Southwest Jiaotong University, Chengdu, China

Alejandro Ramírez-Rojas, Lucía R. Moreno-Torres and Ricardo T. Páez-Hernández
Departamento de Ciencias Básicas, Col. Reynosa, Azcapotzalco, México D. F., México

Israel Reyes-Ramírez
Unidad Profesional Intedisciplinaria en Ingeniería y Tecnologías Avanzadas, México D.F., Mexico

R. Sawires
Department of Geology, Faculty of Science, Assiut University, Assiut, Egypt
Department of Physics, University of Jaén, Jaén, Spain

J.A. Peláez and M.T. García Hernández
Department of Physics, University of Jaén, Jaén, Spain

R.E. Fat-Helbary
Aswan Seismological Center, Aswan, Egypt

Anna Kurzych, Zbigniew Krajewski and Leszek R. Jaroszewicz
Institute of Applied Physics, Military University of Technology, Warsaw, Poland

Krzysztof P. Teisseyre
Institute of Geophysics, Polish Academy of Sciences, Warsaw, Poland

Index

www.ingramcontent.com/pod-product-compliance
Lightning Source LLC
LaVergne TN
LVHW080735210926
848485LV00003B/561